THE I AM MODEL
OF AFRIKAN/AFRICAN/BLACK MALE
IDENTITY DEVELOPMENT

LIONEL MANDY

authorHOUSE®

AuthorHouse™
1663 Liberty Drive, Suite 200
Bloomington, IN 47403
www.authorhouse.com
Phone: 1-800-839-8640

First published by AuthorHouse 7/16/2009

ISBN: 978-1-4389-1429-9 (sc)

Printed in the United States of America
Bloomington, Indiana

The symbols on the cover are adapted from Credo Vusamazulu Mutwa's book, **Indaba, My Children**. *Johannesburg, South Africa, Blue Crane Books. 1964, pages 542-545.*

Contents

DEDICATION

This book is dedicated to:

My ancestors, who have guided me through the rigors of the creation and writing process;

My mother, who advised me to get as much education as I could;

My son, Cabral, for whom I live as a model of what men of Afrikan descent must be and do;

My students at California State University, Long Beach, and at Long Beach City College, who are my teachers;

The faculty and staff at California State University, Long Beach, and at Long Beach City College, who are my mentors and friends;

The healers who have trained and sustained me; and

All people of Afrikan descent, models of the reality and possibilities for all of humanity.

ACKNOWLEDGMENTS

I gratefully acknowledge the efforts of my dissertation committee for guiding me to completion of the initial rendering of this book:

To my chairperson, Dr. Taasogle Daryl Rowe, my gratitude for your patience, the hours of reading, reflecting, and redirecting me during this process (and for lending me those books that I could not otherwise borrow);

To Dr. Shelly Prillerman Harrell, my gratitude for your warm support, energy, and grounding in clinical applications; and

To Dr. Bede Ssendikaddiwa Ssensalo, my gratitude for your contribution to the literature aspects of the dissertation, for clarifying the issue of naming in the Afrikan context, and for your support in the Department of Africana Studies at California State University.

I also thank Dr. Fabunmi Webb-Msemaji for your kindness in lending me your dissertation and a number of books in Black and Afrikan psychology that proved to be indispensable in completing this project.

ABSTRACT

Over the past forty years, Black and Afrikan/African psychologists have developed theories on the Black and Afrikan personality, and on Black and Afrikan male identity development. Some of these theories are centered in Afrikan theory, experience and reality, while other utilize the tenets of mainstream psychology. Models of Afrikan/African-centered male identity development are less fully developed. The I AM Model studies Afrikan-centered psychology to discern its major themes. It then utilizes those themes, as well as themes of identity development found in Ralph Ellison's novel *Invisible Man* to create a new model: the I AM Model of Afrikan male identity development. This new model is compared to current Afrikan-centered models to show how they can be enhanced.

CHAPTER I
The Problem
Introduction: Afrikan American Identity Development
"Is what one learns worth what one forgets?"
(Kane, 1969, p. 31)

Among the many challenges facing persons of Afrikan[1] descent, the process of raising Afrikan boys to manhood is a critical one (Kunjufu, 1985, 1986). In the United States, this task, this journey, has been made more difficult by the structure and function of the society (Bennett, 1988). Persons of Afrikan descen't, many of whose ancestors were brought to these shores as slaves, have yet to be allowed, by persons of European descent[2], to function as full and equal participants in the United States (Kambon, 1998), and have yet to assert their own power to attain a position of equality. The physical, emotional, spiritual, and cultural consequences of this oppression have been equally felt by the men, women and children of Afrikan descent in the United States (Latif & Latif, 1994). Over the course of hundreds of years of enslavement, followed by a century and one half of the appearance of de jure freedom and the reality of de facto subjugation, people of Afrikan descent have struggled to maintain their culture, their hope, their aspirations, and their identity (Bell, 1992). Afrikan males have struggled against a special form of oppression, being whipped, tortured, castrated, lynched and imprisoned by males of European descent bent on maintaining a position of superiority over them (Welsing, 1991). Repeating and reinforcing such treatment for generations has had a devastating effect on young males of Afrikan descent in the United States who aspire to become men in this society (Obadele, 2003). One consequence is the belief that many males of Afrikan descent create

1

facades to hide their real selves and present a strong, confident self to others (Majors & Billson, 1992). This book focuses on males of Afrikan descent, taking as its task the illumination of ways in which the psychological and cultural well-being and the identity development of Afrikan American males can be enhanced as they travel from childhood to adulthood in the United States.

Psychology literature on male identity development

This task is immediately challenged by the relative lack of extant literature devoted to the subject of identity development among males of Afrikan descent. Mainstream (hereinafter "European," "American," "White," or "mainstream") psychology is replete with research, protocols, and theories regarding psychological functioning and identity development (Billson, 1996). These theories and their practical applications have become popular forms of therapy in the United States. However, these theories, and the therapies that are practiced as a result of these theories, have not gained wide acceptance in communities of Afrikan descent (Lee & Bailey, 1998a). People of Afrikan descent in general, and males of Afrikan descent in particular, distrust these theories, the therapy that comes out of them, and the therapists who implement them (Franklin, 1992). The truth of this statement was illustrated in the birth of the Association of Black Psychologists in the late 1960s. The theories and applications produced by members of the Association of Black Psychologists stand as a testament to the failure of mainstream psychology in meeting the needs of persons of Afrikan descent (Baldwin, 1992). In short, if European and American psychology were successful in addressing the emotional/spiritual challenges faced by persons of Afrikan descent, there would exist no need for a psychology (hereinafter referred to as Afrikan or Black psychology) produced by and for persons of Afrikan descent.

There is also a literature on the psychology of adolescents and of adolescence in European and American psychology, a branch of which concerns the advancement of adolescent males to manhood (Sexton-Radek, 2005; Way & Chu, 2004). However, this literature and the research behind it have also been shown to be inadequate in assisting the journey of males of Afrikan descent to manhood (Nobles, 1998). As a result, a literature on Afrikan identity development has begun to emerge within the last two decades. Much has been written in European and American psychology concerning Afrikan men and Afrikan women, but little is available that delineates a path for them to successfully attain adulthood. As an emerging field of study, Afrikan/Black psychology is still in the stage of enunciating and refining its theories. While some work has been done on identity development, that work is in its beginning stages (Jones, 1998). There is little work focused on identity development and the passage of male adolescents of Afrikan descent to manhood (or of female adolescents of Afrikan descent to womanhood). Written work on rites of passage has added vital information and analysis to the study of the transition from childhood to adulthood by Afrikan children (Hare & Hare, 1985; Kunjufu, 1986; Lewis, 1988).

This discussion focuses on the work of Afrikan/Black psychologists in the area of male identity development. This choice is based on their connection with and focus on persons of Afrikan descent. It is in line with a long-standing view that persons of Afrikan descent are best able to assess and address issues faced by their own community (Lee & Bailey, 1998a).

The use of literature to amplify on identity development
A core aspect of this project is the use of a fictional character as an example of and a metaphor for Afrikan male identity development.

Literary critics and analysts have long used the tenets of psychology to analyze and understand the motivations and actions of characters in books and novels, poems and plays (Edel, 1964; Holland, Homan & Paris, 1989). For example, the psychoanalytic framework has long been used to understand and explain the actions of Captain Ahab, the main character in Herman Melville's novel *Moby Dick* (Boker, 1996; Spark, 2001). However, to date literature has not been used to explain or clarify aspects of psychology. While the reasons for this absence are not clear, one can assume that psychology, which fancies itself a science, might not wish to be connected with literature and literary criticism, which are considered to be art rather than science. The arena of Afrikan psychology is not encumbered by these distinctions. Afrikan psychology is holistic: it envelopes all aspects of the lives and cultures of peoples of Afrikan descent. Due to its holistic nature, Afrikan psychology is a continuum of scientific and non-scientific concepts brought together into a whole that works in the service of healing the ills of persons of Afrikan descent (Grills, 2004a; Nobles & Goddard, 1993).

One of the major novels in Afrikan American literature that addresses the concept of identity development is Ralph Ellison's *Invisible Man* (1947). The novel's main character is a young man of Afrikan descent, whose journey in the novel spans his late teenaged and early adult years, during which he is searching for who he is. Therefore, his character is an excellent example of the issues present in identity formation, and can be used to create and discuss theories of identity development.

To begin this endeavor, in Chapter II, mainstream psychology and Black and Afrikan psychology are analyzed. Mainstream psychology has long used a deficit model (Guthrie, 1976) to assert that the intellectual inferiority of persons of Afrikan descent was genetically based. This model and other mainstream psychological theories built up a general societal belief that people of Afrikan descent were socially, intellectually, culturally, and morally inferior to people of European descent (Kambon, 1998; Welsing, 1991). In response to the oppressive nature of these constructs, psychologists and psychiatrists of Afrikan descent developed theories that validated the cultures and personhood of people of Afrikan descent. Eventually this work led these Afrikan psychologists and psychiatrists back to the continent of Afrika to study the philosophical and psychological beliefs found in Afrikan cultures, both ancient and modern. The goal was to derive a set of overarching principles that unifies all the theories in Black and Afrikan psychology. These principles are enunciated and used to frame the analysis of Black and Afrikan psychology in this chapter. Black and Afrikan psychologists also developed theories on identity development which are explained and critiqued. The chapter concludes that there exists a need for a more coherent and comprehensive theory of Afrikan-centered[3] identity development.

Chapter III offers the use of a character derived from Afrikan American literature as a source and metaphor for how such an identity development model might be constructed. The chapter begins by giving a historical picture of the development of Afrikan American literature as we know it through publications that have survived. Starting with its humble beginnings as slave narratives, poems and treatises for the freeing of those of Afrikan descent who were enslaved in the United States, this literature grows more varied in form and content as more authors of Afrikan descent are published. The era in which the novel *Invisible Man* was published by Ralph Ellison (1947) is a juncture in history that is of unique importance to people of Afrikan descent, coming as it does just after the end of World War II and just prior to the advent of the Civil Rights movement. *Invisible Man* is discussed and quoted in detail, with a focus on how the psychology of the character's emerging identity can be seen in the novel. Literary criticism of *Invisible Man* is analyzed to clarify aspects of the novel that will assist in understanding identity development.

Just as Black psychologists and psychiatrists felt the need to create their own theories to explain themselves and their people in psychological terms, so did authors and literary critics of Afrikan descent. They created a school of literary criticism that better fit the needs of artists of Afrikan descent in the United States who were struggling to gain literary respect. The "art for art's sake school" of mainstream literary criticism (Beckson & Ganz, 1989), and the Black aesthetics school of Afrikan American literary criticism (Gayle, 1972) are compared, contrasted, and applied to *Invisible Man*. The chapter concludes with ways in which the novel *Invisible Man* and its main character offer insight into Afrikan-centered psychology and identity development.

Chapter IV brings Afrikan-centered psychology and *Invisible Man* together, showing how the life of the main character in *Invisible Man* fits in with, and offers clarification to Afrikan psychology in general, and to theories of Afrikan male identity development in particular. Nobles and Goddard's (1993) model for prevention of alcohol and substance use and abuse by youth of Afrikan descent is used to frame the discussion of identity development. The main character in *Invisible Man* is used as a test case for the creation of the I AM Model for identity development among males of Afrikan descent in the United States.

Chapter V offers suggestions for further research in this area, and discusses the use of comparative methods, linking "unlikely" disciplines together in the service of theory formation and implementation.

CHAPTER II

'I don't completely *trust anyone, ...not even myself.'*
(Howard-Pitney, 2004, p. 33 [Alex Haley quoting
Malcolm X, *italics in original*]).

Literature Review: Mainstream Psychology and
Afrikan/Black Psychology

Introduction: The Rebirth of Afrikan/Black Psychology

In 1970, Afrikan American psychologist Dr. Joseph White re-ignited
a revolution and symbolically announced the continuation of an
evolution with the publication of his "Toward a Black Psychology"
(White, 1970). Noting the distinction between how persons of
Afrikan descent viewed themselves and how European and American
psychology viewed them, he broke from the victim-focused psychology
that had historically been applied to persons of Afrikan descent. In
stating that persons of Afrikan descent had always been equal members
of the human family, he challenged the notion of the "slave," a creature
created by the environment of her/his circumstances, someone less
than human. This was indeed a revolutionary stance given mainstream
psychology's denigration of people of Afrikan descent. The shockwaves
of this revolution reverberated through the psychological community,
and through communities of Afrikan descent. Women and men in
these Afrikan communities rallied to this cause, as they had been rallying
to similar calls in the Civil Rights, Black Power, human rights, anti-
war, and other movements of that era. This revolt, fueled symbolically
by the picket sign that read "I AM SOMEBODY," and musically by
James Brown's "Say It Loud: I'm Black and I'm Proud," continues to
grow in communities of Afrikan descent around the world, as its effects

sink deeper into a race of people seeking to restore themselves to their proper place in the human family.[4]

White's essay also signaled the continuation of the evolution of Afrikan psychological thought that began eons ago on the continent of Afrika. This evolution is traceable at least as far back as the writings of scholars in Kemet (which the Greeks re-named "Egypt").[5] It provides a basis, a foundation, and a starting point for an analysis of who persons of Afrikan descent are, what connections exist between Afrikans as spiritual beings and as psychological beings, and what implications those seminal conclusions have for who persons of Afrikan descent are today (Akbar, 1994). That basis is a group-centered context which stands in counterpoint to European and American psychology's foundations and systems of analysis (Nobles, 1998).

This discussion will focus exclusively on persons of Afrikan descent in the United States. However, an analysis that correctly describes the basis on which persons of Afrikan descent in the United States stand, may be applicable to persons of Afrikan descent elsewhere. The truth of this statement is left for others to verify.

Applied Psychology: History of European and American Psychology
As William Wundt was constructing what later became known as the first psychology laboratory in Germany in 1879, his class, race, and political contemporaries were preparing to divide up the colonized peoples, nations, and resources of the world in what became known as the Berlin Conference (Franklin & Moss, 2000; Rodney, 1972). It is not by accident that these two seemingly disparate occurrences, political imperialism and the advent of European psychology, took place in the same part of the world during the same period of time. In fact, it could not be otherwise, as it appears that an arrogance and disregard for the rights of other beings led the European elites and their

White American cousins to engage in the conquering and pillage of the world's peoples, nations, languages, cultures, traditions, religions, and spirits, all in the name of profit, plunder, empire, wealth and glory (Karenga, 1993). Coincidentally, as the European powers gathered to divide the world up into parts to exploit and control the persons and the land, the new science that Wundt represented was beginning to divide the theretofore unified human being into parts to study and control the emotions of the person.

Many scholars trace the earliest precedents of modern mainstream psychology back to Greek philosophy, where the issue of whether there was a connection between the body and the mind, or between the mind and the spirit, was debated (James, 1976). After the demise of Greek civilization and its successors in Rome, the Catholic church gained political ascendency in Europe. With the ascendency of the Catholic church as both a political power and the supreme authority on religious matters[6], there came also the ascendency of the concepts of sin, guilt, and alienation which eventually became central to the ideologies of European countries and empires for the next 2000 years (ben-Jochannan, 1972)[7]. It was but a small step from this posture to assigning the lowest place on the Great Chain of Being (Smedley, 1999) to persons of Afrikan descent, just above that of the apes, and far below persons of European descent.

In the nineteenth century, at the height of European colonial dominance of the world, when the field of psychology was born, the concept known as "White supremacy," was formally enunciated. White supremacy is the belief, held by persons of European descent, in the superiority of persons of European descent over all other peoples. It was promoted to justify the oppression of peoples of color by Europeans (Fredrickson, 1981, 1988). This perspective grew into a view of the world where Europeans defined every aspect of life and existence based on their own understanding of the world, with them in the privileged position (Kambon, 1998). It also held that the understandings of the world held by people of other races were irrelevant.

The eugenics movement, which grew to prominence in the nineteenth century, was led by European psychiatrists. They called for the elimination of persons of Afrikan descent, believing that they were valueless for anything other than manual labor (Citizens Commission on Human Rights [CCHR], 1995a). Margaret Sanger, founder of Planned Parenthood and an American eugenicist, created a plan to hire black ministers to preach to Black people about the use of sterilization of women of Afrikan descent as a solution to poverty (CCHR, 1995a).

In the early part of the twentieth century, psychometrics replaced eugenics theories as a way of categorizing and inferiorizing persons of Afrikan descent. Intelligence Quotient (IQ) testing scores were selectively interpreted to reinforce the inferiority of persons of Afrikan descent as compared to persons of European descent (Smedley, 1999). It was therefore not surprising when Black prisoners were used for psychosurgery experiments in Louisiana in the 1950s: "...it was 'cheaper to use Niggers than cats because they were everywhere and cheap experimental animals'" (CCHR, 1995b, p. 9). Herrnstein and Murray's *The Bell Curve* (1994) reinforced the thesis that people of Afrikan descent are, as a race, genetically inferior to persons of European descent. Hilliard (1991) and other psychologists of Afrikan descent have shown the speciousness of all standardized tests of intelligence as measures of the intelligence of people of Afrikan descent and as predictors for academic success.

Lott's (2002) description of how those with wealth separate, exclude, devalue, discount, and distance themselves from the poor on both institutional and interpersonal levels reflects how those with white supremacist privilege treat those without it. "Psychological theories are preoccupied with people who are like those who construct the theories" (Lott, 2002, p. 101). In line with this view, mainstream clinical psychology focuses primarily on the individual as independent from the social system in which he or she lives (Holdstock, 2000). Helms (1994) states that a hierarchy exists in American society, on virtually every level, where Whites are at the top, followed by Asians, Native Americans, and persons of Afrikan descent, in that order. Sizemore's (1972) conclusion that western modes of knowing leave out important aspects of reality: "insight, understanding, empathy, intuition, prehension, and common sense," adds breadth and depth to this analysis (p. 149).

Radical psychologists conclude that political and psychological oppression coexist and are created in concert (Watts, Griffith, & Abdul-Adil, 1999). Acknowledging that they benefit by the same system that they are criticizing, these (mostly White) radical psychologists conclude that the group in power (also of European descent) must give power and control to the powerless because it is ethically responsible to do so (Prilleltensky & Gonick, 1994). Bartky (1990) notes that within each culture, the psychological and the political are intertwined and inseparable. Control over the disenfranchised, such as persons of Afrikan descent, was delegated by those in power to academics in the field of social science, who were to erect theories, policies, and programs to institutionalize the suppression of the disenfranchised. Bartky (1990) concludes that the function of academics was to separate political issues from social science, thereby controlling the pace and direction of social change. In aligning themselves with the power sectors of society, mainstream psychologists- researchers, therapists, and professors- became "highly sophisticated agents of social control" (p. 162). Fetterman, Kaftarian, & Wandersman (1996) state that since the dominant group makes and enforces rules and standards for all groups, under such circumstances, "differences become deficits" (Fetterman, et al., 1996, p. 24) and contributions made by peoples of color are dismissed. The conclusions reached as a result of this radical critique of mainstream psychology is that a psychology focused on liberating subjected groups based on moral values must be encouraged (Bartky, 1990). However, no such psychology has emerged to date.

King (1980) noted that none of the seven models of psychopathology currently used in mainstream mental health (heredity, developmental, existential, learning/conditioning, neurophysiologic, ecological or interactive) take either class or culture into account. All seven isolate the person from the environment in which they operate and to which they must return after receiving assistance. The implication is that if one is oppressed by society, to assist the person without "fixing" the environment is of little help in restoring the person to well-being. Allen (2001) points out that people of Afrikan descent worldwide have suffered similar types of subordination: elimination of personal identities; separation of language groups and forced acquisition of a foreign language; inculcation of fear through violence; devastation of the family unit; sexual abuse of Afrikan women; creation of a color caste system; forced religious conversion to Europeanized forms of Christianity; and an ongoing propagation of the myth of the Afrikan as a savage.

That system remains in effect to this day. In referring to a study on whiteness, Austin (2001) called making racism a white person's problem "a brilliant strategy" (p. 4). In fact, it is more than a strategy: it is a reality. Racism is the carrying out of oppression by a person or group of people with the power to oppress. People of Afrikan descent have been the victims of racism created and maintained by people of European descent throughout the world (Welsing, 1991). As such, racism is not so much as strategy as it is a system of worldwide domination of peoples of color based on race (Welsing, 1991). White (1970) clearly stated that the theories developed by White people to explain white behavior were inadequate at best, and largely useless in explaining the behavior of persons of Afrikan descent. Asante (1980) went on to posit that mainstream psychology assists in the perpetuation of the system that oppresses persons of Afrikan descent in America. Bulhan (1985) takes the discipline of psychology to task for not revealing the greed and violence that Europeans used to subdue the rest of the world, as well as the prejudice and complicity of many of the pioneers of the discipline in its establishment in Europe. Carruthers (1996) stated that mainstream scientific methodology is of little use to oppressed persons because it is not only used to control the powerless, but was created for that purpose. "The original assumptions *are* oppression, suppression, and repression" (Carruthers, 1996, p. 188, [*italics in original*]). As a consequence of this system and the psychology that supports it, people of Afrikan descent are systematically denied participation in decisions that affect their lives. And when a focused psychological study of persons of Afrikan descent is undertaken, elements of the mainstream academic community chastise these scholars by accusing them of a form of racism (Tucker & Herman, 2002). Wilson (1993) states that given a racist system, psychological diagnosis and treatment of oppressed persons become political acts with political goals and consequences. He goes on to note that such a system cannot provide adequate psychological care for persons of Afrikan descent who, of necessity, are driven by a desire for liberation from subordination (Wilson, 1993). He concludes that for persons of Afrikan descent to be controlled without being physically imprisoned as a people, they must be controlled psychologically, or, as he puts it, must be out of their minds, and kept out of their minds (Wilson, 1993). At the United Nations conference on racism, held in Azania[8] in 2001, the American Psychological Association placed a declaration into the record. It called for the acknowledgment of racism in the world, and for the establishment of research to study the problem. But it offered no solutions (Murray, 2002).

One result of centuries of racist indoctrination is exemplified by a Jewish civil rights activist's recent anecdote about his grandmother, who suffered from Alzheimer's disease:

> She could not remember how to feed herself. She could not go to the bathroom by herself. She could not recognize a glass of water for what it was. But she could recognize a nigger. America had seen to that, and no disease could strip her of that memory. (Wise, 1999, p. 3)

It is this history of oppression toward persons of Afrikan descent, and the responses of persons of Afrikan descent to that treatment, that brought forth the need for a psychology that reframed the problems faced by persons of Afrikan descent, and the solutions to those problems. This treatise asserts that such a new paradigm, once created, must be elaborated so that it is of practical use in healing long-standing wounds suffered by people of Afrikan descent.

The Birth of Afrikan/Black Psychology

A history of the origins of Afrikan/Black psychology that flows in chronological order would, on the surface, seem to make the most sense. However, given the history of anti-Afrikan/anti-Black supremacy domination by persons of European and Asian descent in the writing and understanding of history in general, and of psychology in particular (Kambon, 1998), such a chronology does not best explicate the circuitous path taken in the formation of Afrikan/Black psychology.[9] The better course is to view Afrikan/Black psychology as it evolved into the consciousness of its practitioners. Taking this path, the rebirth of the field preceded knowledge of its birth. However, the ancient core of Afrikan psychological philosophy and theory forms the basis for this rebirth. Therefore, this core will be discussed first to preserve the position of ancient Afrikan ancestors, who brought this field of inquiry into existence (Akbar, 1985a).

The oldest extant, documented evidence for an Afrikan psychological framework is found in Kemet on the Hu-Nefer papyrus from the reign of Pharaoh Seti I, also known as the Papyrus of ANI (Budge, 1915). This scroll depicts a scene that has been named "Psychostasia" (Nobles, 1986, inside cover). This scene depicts the Ka (the spirit or personality) of Hu-Nefer, scribe and keeper of the cattle for Seti I, reciting from *The Book of Coming Forth by Light*. The name Hu-Nefer (Gadalla, 2001) represents a mature (nefer) authoritative utterance (hu). This recitation takes place in the presence of Heru and Tehuti, representing the will and the way, Ma-at, representing truth, justice, and balance, Ammit representing a person with no consciousness, Ausar, judge of all, and others (Gadalla, 2001). This depiction represents the weighing of the conscience of Ani, or Hu-Nefer. It is symbolic of the union and interdependency of the physical, emotional, mental, spiritual and universal aspects in human functioning. In other words, right conduct on all levels of personal and community life is essential to achieving human balance or normality.

The civilization constructed in Kemet has been called the culmination of all prior Afrikan civilizations by many Afrikan scholars (ben-Jochannan, 1971; Jackson, J. G., 1970; Williams, 1987). These scholars agree, and present-day Afrikan psychologists concur, that the precepts found in Kemet spread from the interior of Afrika to Kemet, and express an ethos that was and is found throughout the continent (Nobles, 1986). These precepts on what constitutes psychological balance have come to us in languages other than those of Afrikans, but the concepts remain unaltered. The wisdom of Kemet comes to us in symbols (the writings in the language then called Mdu Ntr, which is now known as hieroglyphics, are a series of symbols) and myths (Carruthers, 1995). Akbar (1985a) discussed the seven aspects of the soul, as they were known in Kemet, which was synonymous with the self. Akbar also notes that in Kemet, man and woman were metaphors for higher truths. Kemite psychology was metaphysically based, with myths as metaphors or symbols for ideas and emotions. Chandler (1999) notes that mythologies, such as those found in ancient Kemet, became a vehicle for the transmission of truth, rather than merely of facts.

The Kemite core principles of psychological balance are found in what has been called Hermetic philosophy, named after Hermes Trismegistus, the Greek name ascribed to the Kemite god Troth, or Tehuti (Gadalla, 2001). These principles are found in a book entitled *The Kybalion* (Three Initiates, 1940). Concerning the Kemites, the authors of *The Kybalion* state: "Nor must it be supposed that they were ignorant of the so-called modern discoveries in psychology-- on the contrary, the Egyptians were especially skilled in the science of Psychology" (Three Initiates, 1940, p. 44). This psychology consisted of a combination of what is now known as metaphysics, as well as what is now known as psychology (Three Initiates, 1940).

The first of these principles is: "The Principle of Mentalism: THE ALL is MIND; The Universe is Mental."" (Three Initiates, 1940, p. 26). This "ALL" is the sum total of the universe, and all within as well as beyond it, including all matter, all energy, and all that human beings are able to perceive. Chandler (1999) says that we call THE ALL "God." "MIND" includes the concept of "Mental Transmutation," the changing of one mental state into others (Three Initiates, 1940, pp. 48-49). These changes can be affected by the person who desires the change, and works to achieve it. By doing so, the person will escape the challenges presented by life on the lower or everyday levels of life. In this sense, humans reflect the ALL in our ability to create using our minds. This "ALL" is also known as "AN UNIVERSAL, INFINITE, LIVING MIND" (Three Initiates, 1940, p. 27). The principle of THE ALL also incorporates the idea that change is the only permanent reality. Change is a manifestation of an underlying power which is called "THE ALL." This unknowable, infinite, unchanging MIND or ALL is also called "SPIRIT" (Three Initiates, 1940, p. 63). Spirit, in this sense, is a property possessed by individuals, and by the collective, the community. According to the Buddha, change is a basic feature of nature (Chandler, 1999). Viewed from this principle, resisting change is a cause of most psychological disease.

The second principle in *The Kybalion* is: "The Principle of Correspondence: As above, so below; as below, so above."" (Three

Initiates, 1940, p. 28). This principle indicates that the material, mental, and spiritual universes mirror each other, and that all are governed by the same universal laws (Three Initiates, 1940). By knowing the essence of the operation of things on the material level or plane, one is able to understand the operation of things on the mental, spiritual or universal level or plane. This insight allows us to observe the known and deduce the unknown from it. An example of this axiom is the correspondence between the way that the planets in our solar system revolve around the sun, and the smaller "universe" of protons, electrons, and neutrons revolving around the nucleus of an atom (Chandler, 1999). It is this principle to which Akbar (1985a) was pointing when he described the seven aspects of the soul in ancient Kemet. These seven aspects are symbols for higher truths, and the soul is the center of a human being as the sun is the center of the solar system. And like the harmony that exists when all the planets are in alignment, so it is with human beings when their emotions are in alignment with their thoughts.

The third principle of *The Kybalion* is: "The Principle of Vibration: Nothing rests; everything moves; everything vibrates.'" (Three Initiates, 1940, p. 30). This principle means that the differences we experience between things: "Matter, Energy, Mind, and even Spirit" are the differences in the rates at which they vibrate (Three Initiates, 1940, p. 30). The higher the rate of vibration, the higher the form. The lowest form is matter, which vibrates at the lowest rate, a rate at which we as human beings can perceive it. Hence a rock is vibrating at a rate slow enough that we can perceive it as a discrete form of matter. Wind operates at a higher vibration. We cannot see the wind vibrating. Rather, we perceive it indirectly, either by feeling it, or by seeing the effect of its vibration on trees, clouds, or other objects. The highest form of vibration is spirit, which vibrates so fast that it appears to not be moving at all (Three Initiates, 1940). An electric fan illustrates this principle. When it is at rest, we can see each of the blades of the fan. As it begins to move (vibrate faster), we continue see the blades, but lose some of the clarity of the shape of each blade. As the fan moves still faster, the spaces between the blades are now filled in, and the fan appears to be solid. At this point, the fan does not appear, visually, to be moving at all. The importance of this principle is that all feelings, thoughts and emotions are accompanied by specific vibrations. These vibrations are sensed by others, though they may not be seen by the observer. However, they may be felt, which appears to indicate that differing emotions vibrate at differing frequencies.

The fourth principle in *The Kybalion* is: "The Principle of Polarity: Everything is Dual; everything has poles; everything has its pair of opposites; like and unlike are the same; opposites are identical in nature, but different in degree; extremes meet; all truths are but half-truths; all paradoxes may be reconciled'" (Three Initiates, 1940, p. 32). Utilizing this principle, one can see that love and hate are two poles of the same emotion. Many of the core principles of modern society operate by distinguishing two poles of an issue rather than recognizing that they are two poles of the same issue. Hence, this society focuses on distinctions such as good and bad, legal and illegal, intelligent and unintelligent, rich and poor, rather than seeing their similarities. Combining this principle with the Principle of Vibration, the changing of emotions involves changing the vibration of hate to one of love along the pole of hate-love, both in oneself and in others (Three Initiates, 1940). Other ranges of emotions, such as from fear to courage, can be addressed in similar fashion. The change of vibration to move from one pole of emotions can be done by individuals and communities. In addition, the vibration given off by one individual or community can effect or change the vibration of another individual, groups of individuals, a community, or a group of communities. Malcolm X gave off a vibration of confidence in the face of opposition, whether he was on a television show or was speaking on a street corner. His confidence instilled confidence in his audience, which resulted in the mobilization of a community of persons of Afrikan descent for change.

The fifth principle of *The Kybalion* is: "The Principle of Rhythm: Everything flows, out and in; everything has its tides; all things rise and fall; the pendulum-swing manifests in everything; the measure of the swing to the right is the measure of the swing to the left; rhythm compensates.'" (Three Initiates, 1940, p. 35). This rhythm is one of action and reaction, rise and fall (Three Initiates, 1940). Hence, persons and plants rise (are born) and fall (die), as do nations, ideologies, and the tides in the oceans. Similarly, we inhale and exhale in a rhythm. The sun and the moon appear to rise, and then fall, to rise again. This principle is also applied to the movement between lower consciousness and higher consciousness. It is the mastery of the ability to neutralize the swing of the pendulum that allows a person or a people to attain self-mastery. In this state, depression need not follow enthusiasm, and fear need not follow courage (Three Initiates, 1940). This principle is currently encapsulated in the saying: "What goes around, comes around." Chandler (1999) applies this principle to people of Afrikan descent. Once they were powerful and wealthy, and now they are oppressed and poor. This perspective implies that the pendulum will again swing in a positive direction for Afrikan peoples in the future.

The sixth principle from *The Kybalion* is: "The Principle of

Cause and Effect: Every Cause has its Effect; every Effect has its Cause; everything happens according to law; Chance is but a name for Law not recognized; there are many planes of causation, but nothing escapes the Law.'" (Three Initiates, 1940, p. 38). This principle indicates that there are no accidents in life; nothing occurs by chance. Chance is an event whose cause has not been recognized. And each effect, such as a current state of affairs, is made up of millions of causes. An example of this is seen in each human being who is alive at present. Each of us who is alive today is an effect of each of our ancestors, millions in number, each of whom gave birth to a child, who form our direct ancestral lineage. This principle also indicates that many human beings are not acting on their own desires, but are operating on the feelings and emotions created by others, as well as on the thoughts and customs created by others. For example, many people in the United States believe that a certain religious/ethnic group of people are terrorists based on images that the electronic media project about people who are opposed to the policies of the government of the United States. This principle is also seen in the relation between the phases of the moon and the height of ocean tides. Similarly, the closeness of the sun to the earth, the cause, has an effect on the temperature we experience on the surface of the earth.

The seventh principle in *The Kybalion* is: "The Principle of Gender: Gender is in everything; everything has its Masculine and Feminine Principles; Gender manifests on all planes.'" (Three Initiates, 1940, p. 39). This principle indicates that every being has both male and female principles, to varying degrees, within them, regardless of which gender appears to be dominant physically. In this framework, masculine energy is the energy that initiates creation, while the feminine energy actually creates. These concepts of creation have a much broader connotation than sexual intercourse, which is a manifestation of the principle on the material, physical level of life. This principle also operates at the mental and spiritual levels, though we are often not aware of its operation. The masculine principle is seen in those who can impress their beliefs on others, a quality also known as charisma. A leader, who is expressing the masculine principle, requires followers, the feminine principle, in order to bring their visions to fruition. This principle is also seen in the early stages of the life of a human fetus. In the early stages, the fetus has the ability to become either a male or a female human being. Hence, regardless of which gender it becomes, the fetus retains the qualities of the other gender. Chandler (1999) posits that the denial of the feminine aspect of creation and the assertion of the masculine aspect has resulted in destruction in every Western culture as well as in every culture influenced by the West. This idea is most readily seen in preference for the masculine spiritual principles and the minimization of female spiritual principles, or in the preference for gods over goddesses.

These (Hermetic) Tehutian principles constitute universal attributes that are found in all matter including human beings. This is why the Kemites were called the "original psychologists" (Three Initiates, 1940, p. 43). Using these principles, a person who was out of balance, out of harmony with the universe around them, was taught to use their will to change their vibration, mental state, or state of consciousness. This was done by focusing on the opposite pole to the one currently experienced (for example, focusing on happiness when one was depressed). This would cause the person's vibration to change from negative to positive, and the person would experience themselves as moving toward the positive vibration of the emotion. The ability to do this consistently would build character and eventually would result in mastery on all levels of being. When one achieved a positive polarity by focusing on the positive pole of an issue, it was possible to change the emotion they experienced, to alter how they felt, and to strengthen their character. Having achieved balance, the Law of Neutralization became important. This law says that human beings are capable of controlling the operation of the movement between the poles, so that we can maintain ourselves on the positive pole, thus enjoying happiness, for example. The person maintains a positive vibration, and gains control over the issue, becoming an agent causing, rather receiving. In the above example, a person can make herself or himself happy and remain happy, rather than becoming depressed and remaining depressed by the actions of others.

Using these principles, our Kemite ancestral psychologists' brought their own mental vibration to the opposite pole of the emotions of their patient', and moved the patient' from a negative toward a positive state of emotions (Three Initiates, 1940).[10] The goal of all Kemite healing was Maat: "harmony, balance and equilibrium" (Gadalla, 2001, p. 41). The principle of Polarity contains within it the Principle of Compensation (Three Initiates, 1940, p. 167). This principle states that one who experiences a strong positive emotion is likely to experience an equally strong negative emotion. Hence, the achievement of Maat is the balance that comes from maintaining a positive vibration so as to achieve mastery over emotions. These seven Tehutian principles constitute the core laws and principles of ancient Afrikan psychology. They are a collection and organization of principles deduced by Afrikan peoples and transmitted to each other over the eons of Afrikan existence prior to the creation of the civilization of Kemet, and recorded (written) for history by the Kemites.

With the demise of the empire of Kemet, the contributions that the Kemites made were absorbed, claimed, pillaged, and retranslated by the Hittites, Macedonians, Greeks, Romans, Arabs, and the British (James, 1976; Williams, 1987). The mysteries systems, as the compilation of wisdom in Kemet was known, was transformed and lost, absorbed and diffused by cultures around the world over time (Chandler, 1999). Bynum (1999) chronicles the diffusion and alteration of this knowledge into Europe. He notes that the Greeks, who are credited with the creation of psychology, took parts of the Tehutian principles and brought them back to Greece. It is Bynum's contention that the Greeks either did not understand Kemite psychology, or that they did understand it and chose to ignore it in its totality. In the last two centuries, scholars have sought to resurrect and understand the knowledge left by the scholars of Kemet in its original form. This work proceeds from two sources: European interest in Egypt, and Afrikan interest in Kemet. European scholars such as Budge (1915), Petrie (1909), Gardiner (1927), and others used the deciphering of the Mdu Ntr scripts found on the Rosetta Stone in an attempt to claim Egypt as the first great European civilization. Afrikan scholars such as ben Jochannan (1981), Carruthers (1984), and Diop (1974, 1990), have studied the writings in their original form, the European translations, the history of Kemet, and the ethos of Afrikan cultures, to correct history and credit the continent of Afrika with giving birth to and populating Kemite civilization. These Afrikan scholars emerged from mainstream history and sought to understand and reclaim their own history in Afrika. In Kemet, these Afrikan scholars saw an Afrikan culture, their culture, which flourished prior to being attacked by other cultures (Clegg & Ahmed, 1999; Obenga, 2004). Afrikan scholars in the United States, where the dominant culture is oppressing people of Afrikan descent, are part of this process of reclaiming and reframing

Afrikan culture (Parker, 1918/1981; Rashidi, 1992). Kemet became their paradigm. The historical documentation of the existence of the study of the soul, or mind (soul or mind are translations of the Greek term that makes up the modern-day term "psychology") (Merriam-Webster's Collegiate Dictionary, 2001), gave twentieth century Afrikan scholars a renewed and clear understanding of what is now called Black and/or Afrikan psychology (Akbar, 1985a). This reconnection is further elucidated below.

The process of resurrecting Kemet and its contributions did not end there. Afrikan scholars have shown direct connections between Kemet and modern peoples found on the Afrikan continent. Diop (1980) and Obenga (1997) proved the direct linguistic connection between Mdu Ntr and West Afrikan languages, and particularly Wolof, which is currently spoken in Senegal and the Gambia. Subsequent research proved that all Afrikan languages are descended from a common root, the Kemite language called Mdu Ntr (Wimby, 1986). Bynum (1999) describes the diffusion of the collective Afrikan consciousness from its Afrikan roots, through its study and expression in Kemet, back to the various peoples of Afrika, where it can be found today. Gadalla (1999) explains how the people of Kemet moved to western and southern Afrika as Kemet declined, bringing with them their religious beliefs, their social and political structures, and their organization of labor. Gadalla also shows the close comparison between the beliefs and practices of healers in ancient Kemet with healers in modern-day, traditional Afrikan societies. This perspective reflects the views of earlier Afrikan scholars (ben-Jochannan, 1971; Clarke, 1999; Williams, 1987). This diffusion from Kemet throughout the continent of Afrika explains how and why West Afrikans forcibly taken from Afrika and transported to the western hemisphere can retain the essences of Kemite society though they are physically removed from Afrika (Clarke, 1991; Karenga & Carruthers, 1986).

The Rebirth of Afrikan/Black Psychology: Background

A confluence of streams resulted in the rebirth of Afrikan/ Black psychology. One stream was the ongoing struggle of Afrikan people in the western hemisphere in general, and in the United States in particular, for freedom and equality (Harding, 1981). A second stream was the cry by the peoples of Afrika for an end of colonial domination of the Afrikan continent by European countries (Davidson, 1964; Rodney, 1972). These two streams came together in the 1950s. In the United States, the stream flowed into the Civil Rights movement, which overflowed into the Black Power movement in the 1960s and 1970s. Simultaneously, on the Afrikan continent, the peoples of various nations were throwing off the chains of colonialism, and gaining political independence. The undercurrent to both of these streams was the re-emergence of racial pride and peoplehood that had long-standing and deep-seated roots.

On the Afrikan continent, this quest for freedom resulted in the works of Frantz Fanon, an author of Afrikan descent, works about the psychological challenges faced by Afrikan peoples in search of their freedom (Fanon, 1963, 1965, 1967a, 1967b). Fanon's theories on the need for the emergence of a new Afrikan man and woman inspired his compatriots in Algeria, from whence he wrote. His analyses also inspired members of the emergent Black Power movement in the United States. For example, his book *The Wretched of the Earth* was mandatory reading for all incoming members of the Black Panther Party for Self-Defense (Seale, 1968).

Fanon was a psychiatrist. He was born in Martinique trained in France. He was stationed in Algeria as psychiatrist in a mental institution operated by the French during the Algerian revolution against France in the 1950s (Caute, 1970). His central theory was that oppressed people who reacted to their oppression by non-cooperation or by violence against their oppressors reflected an emergent consciousness that could be used to generate organized action against oppression (Banks, 1979). Bartky (1990) cites Fanon for the truism that one could be oppressed psychologically, as well as economically, legally, or physically. To be so was to be "weighed down in your mind" (Bartky, 1990, p. 154). In so doing, according to Fanon, the psychologically oppressed person become their own oppressor by mentally subverting their own self-esteem and internalizing their own inferiority (Fanon, 1963). Fanon's thesis has two parts. One part discusses the depersonalization that persons of Afrikan descent suffered from their experiences with the oppression of colonialism (Fanon, 1967a). In this respect, he respected Jung, but criticized Jung's theory of collective unconscious for being racist (Bulhan, 1980). The other part was the freeing up of a hidden creativity that came with struggling for liberation and freedom. This struggle resulted in reconnecting with an Afrikan culture or cultures that nurtured them (Fanon, 1967b). As part of the process, internalized violence (the violence of persons of Afrikan descent against each other) is turned outward against sources of oppression. By doing so, these persons of Afrikan descent became both the subjects and objects of history (Bulhan, 1980). Fanon grew to believe that he should not provide therapy for men and women of Afrikan descent so that they could adjust to living in a racist society (Dixon & King, 1980). Fanon's writings inspired a generation of persons of Afrikan descent in America (Newton, 1973). This inspiration came in the form of a political analysis that defined the position of persons of Afrikan descent in the

United States as members of an internal colony, which was seen as an extension of the enslavement and the plantation systems (Nkrumah, 1965; Obadele, 1975, 1997; Peery, 1975). Psychologists of Afrikan descent in the United States, who took this political and ideological position, developed the same analysis (King, Dixon & Nobles, 1976). The theories that they developed which evolved from this position came to be known as the reformist (Karenga, 2002) or Black psychology position (Kambon, 1998).

Karenga (1993) suggested that Black psychology was divided into three differing schools of thought: the traditional, reformist, and radical schools.

Black psychology: The traditional school. Psychologists in the traditional school offered no alternatives to mainstream psychology. Rather they posited that telling persons of European descent of the oppression they had perpetrated on persons of Afrikan descent would change the attitudes and actions of the descendants of those European oppressors (Clark, 1965; Grier & Cobbs, 1968). Psychologists in the reformist school stood for change, and advocated a Black psychology, but did so within an American context that took cognizance of the benefits that would accrue for those of Afrikan descent and of European descent (Karenga, 1993). The thrust of this school was to add the information from Black psychology onto mainstream psychology, thus further humanizing it mainstream psychology. Psychologists in the radical school focused exclusively on persons of Afrikan descent. Members of this school base their psychology on a worldview that is centered in Afrika, not Europe. They are scholar activists, working toward the mental health of persons of Afrikan descent through political and cultural struggle, as well as via the practice of psychology

(Karenga, 1993).

Black psychology: The reformist school. White (1970) raised the clarion call for the formation of a Black psychology, charging that mainstream psychology's theories can explain persons of European descent, but not those of Afrikan descent. He advocated the creation of a Black psychology that would take those aspects of mainstream psychology that functioned well for persons of Afrikan descent, and add to them theories based on understanding of the lives of Afrikan peoples. It is worth noting that White first published his call in Ebony, a magazine popular in the Afrikan community at that time. White has said that he did this so that the Afrikan community at large could join in the work of reframing psychology so that it was of benefit to and for them. White's sentiments echoed those of Clark (1963), who noted that persons of Afrikan descent fluctuated between two fantasies. The first was the fantasy of accommodation or acceptance, the belief held by some persons of Afrikan descent that they were better than others of their race. The second was the fantasy of militancy, the belief in a separate black nation, based on the conclusion that total integration of persons of Afrikan and European descent was not possible.

The newly formed Association of Black Psychologists called for a moratorium on research on racial differences and a boycott on intelligence testing due to the racial bias found in those areas (Wilcox, 1971). This call by the Association of Black Psychologists has a solid basis in research (Hilliard, 1991). Bianchi, Zea, Belgrave, and Echeverry (2002) noted that Afrikan Brazilian men who embraced anti-Afrikan/anti-Black values and standards had a more negative attitude toward themselves than did those who had a positive Black racial identity. These consistencies are echoed by Franklin (1992), who notes that American society denies access to opportunities for males of Afrikan descent via rules and codes, thereby creating in them "a sense of invisibility'" (p. 353). Holdstock (2002) notes a similar pattern in the under-representation of Afrikan indigenous, cultural, or folk psychology in research and publications. Responses to this "invisibility" by psychologists are made to empower individuals of Afrikan descent to gain greater control over themselves, and to build skills with which to operate more effectively in the United States (Allen-Meares & Burman, 1995).

At first glance, Afrikan or Black existential thought appeared to offer clarification or a fuller explanation of the issues facing persons of Afrikan descent. However, a review of the first texts written on the subject (Gordon, 1995, 1997, 2000) shows that the authors adapt European conceptualizations of existentialism and apply them to the situations found in persons of Afrikan descent. For example, this Afrikan existential theory identifies the Afrikan victims of racism and enslavement as possessing the absence of the status of "Other" (Gordon, 2000, p. 61). This theory comes from the worldview of the European as master and Afrikan as a slave, or a nonbeing in Fanon's (1967a) parlance. This perspective is little different in focus from those of Jung and other existential philosophers and psychologists who Fanon and others criticized (Bulhan, 1980). And this perspective contains the same limitations: it is framed on a European-centered worldview that precludes realistic discussion of the lives, values, and spiritual origins of persons of Afrikan descent. Therefore, these Afrikan existentialist theories are also reformist in nature.

Azibo provides a summary of what Karenga calls reformist theories, and what Azibo (1988) calls negativist theories. In Azibo's view, these theories are not beneficial to persons of Afrikan descent. Furthermore, Azibo asserts that these theories are actually harmful to persons of Afrikan descent because they use comparative assumptions and data (comparing persons of Afrikan descent to persons of European descent). Finally, these theorists do not utilize the perspective of persons of Afrikan descent in their theories. The result is that the entirety of their analyses come in the context of experiences lived under a culture of anti-Afrikan/anti-Black supremacy.

Sellers, Chavous and Cooke (1998) noted that a nationalist ideology, one that stresses the uniqueness of being of Afrikan descent, requires that Afrikans in America be in control with "minimal input" from other groups (p. 13). The radical school rejects even minimal input" from other groups regarding the destiny of persons of Afrikan descent. Baldwin (1992) criticized Black psychologists, accusing them of colluding in the oppression of persons of Afrikan descent, rather than in their positive mental health. Jackson & Stewart (2002) clarify the differences between those in the reformist and radical schools. They see radical theorists as developmental, focusing on normal functioning, and on the collective nature of personality. By contrast, reformists are focused on transformation and on the individual. In addition, radical theorists emphasize spirituality as a key component of identity (Jackson & Stewart, 2002).

Black psychology: The radical school. The radical school derived much of its impetus from a paradigmatic shift in understanding the place of persons of Afrikan descent in the world. Key to this shift were the Civil Rights and Black Power movements in the United States, which inspired scholars of Afrikan descent to take a new look at their beliefs and analyses (Carmichael & Hamilton, 1967). Out of this paradigm shift, and following the birth of the Association of Black Psychologists, Nobles (1972, 1980) developed the concept of the "extended self." By that he meant that persons of Afrikan descent are defined by, and responsible to the group from which they spring. This is an extension of the Tehutian principle of cause and effect, in that what happens to one person of Afrikan descent effects all members of the group, village, nation, or community. Clark, McGee, Nobles and Akbar (1976) expanded the definition of an African psychology by stating that Afrikans were the first race to inhabit the earth, that the increased amounts of melanin found in persons of Afrikan descent result in increased emotional arousal, that Afrikan intelligence is made up of a combination of psychic abilities and consciousness or self knowledge, and that the Afrikan personality is an extended self, restating Nobles' concept. This definition of Afrikan intelligence aligns with the Tehutian principle of polarity. In western conceptualizations, psychology and psychic abilities are dichotomized, whereas in this Afrikan definition, the poles, while seeming to be opposites, are two aspects of the larger reality of human ability. Asante (1980) developed the theory of "Afrocentricity," defined as "the belief in the centrality of Africans in post modern history. It is our story, our mythology, our creative motif and our ethos exemplifying our collective will" (Asante, 1980, pp. 9-10). Central to Afrocentricity is "Njia," which Asante (1980) defines as "the collective expression of the Afrocentric worldview based in the historical experience of African people" (p. 26). Njia expresses itself as a universal Afrikan consciousness, a connection to the past and the future of the persons and cultures of Afrika. In this respect, Asante's work presages that of Bynum (1999), who proposes an Afrikan unconscious as the central to the lives of all persons of Afrikan descent. Asante (1980) suggests that an Afrikan psychology is concerned with form, feeling and rhythm as key components of an Afrikan aesthetic (p. 95). These components echo the Tehutian principles of rhythm and polarity, uniting form and feeling, seeming opposites, into one whole: humanity. Long (1993) points out that

Afrocentricity in the United States requires a belief in an Afrikan cultural system into which Afrikan and Afrikan American realities are placed, based on values that grow out of Afrikan American experience. Clarke (1994) preferred to call Afrikan-centered thought "Africancentricity" rather than "Afrocentricity" in order to clearly identify its source. He defined "Africancentricity" as "any sincere effort on the part of African people . . . to regain what slavery and colonialism took away and to restore the nation as you originally conceived it to be" (Clarke, 1994, p. 121). Clarke is enunciating the Tehutian principle of polarity, in which people of Afrikan descent should move from the pole of oppression back to the pole of freedom, honor, and humanity.

Worldview as a paradigm. Worldview analysis was a point of demarcation between Black and Afrikan psychologists, and therefore deserves special attention. Smedley (1999) defined race as a worldview imposed by those of European descent on all of the peoples of the world. On the other hand, Asante's Afrocentricity places persons of Afrikan descent at the center of analysis. In other words, his worldview is "Afrikan centered" or Afrocentric (Asante, 1980). Baldwin (1992) forthrightly stated that "the basic nature of the European Worldview is diametrically opposed to the psychological development, survival, and liberation of Black people" (p. 54). An anthropologist describes the Afrikan worldview as a personal self reflecting and mirroring the order of the universe (Holdstock, 2000). This is an expression of the Tehutian principle of correspondence: as above, so below. Allen (2001) sees the Afrikan American worldview as more collective, less competitive, group oriented, and less material than the European American worldview. Nobles (1976) adds that the Afrikan worldview is guided by the survival of the collective as well as harmony with nature.

Diop (1991) dichotomized the world's peoples into two cradles, one European and one Afrikan. He stated that the European cradle was patrilineal, polytheistic, individualistic, aggressive and racially oriented socially, and consisted primarily of the economics, lifestyle, and culture of nomadic hunters. The Afrikan cradle was characterized as matrilineal, monotheistic, communal and collective, congenial and non-racial in social relations, and complementary on all other levels (Diop, as cited in Kambon, 1992, p. 9). Diop's theory reflects the Tehutian principle of polarity, with the two cradles representing the two poles. The European traits that Diop describes become standardized in European philosophy. They are expressed in such dichotomies as strong versus weak, good versus bad, superior and inferior, and black and white (Kambon, 1992).

Azibo (1996a) analyzes European psychology from an Afrikan-centered perspective, a perspective he shares with Wright (1984), Baldwin (1992), and others, who have unequivocally stated that western society in general has as its goal and practice the oppression of persons of Afrikan descent. Kambon, also known as Baldwin (Baldwin, 1992) states that the Afrikan worldview consists of "principles of inclusiveness and synthesis, cooperation and collective responsibility, groupness, sameness and commonality, and spirituality (p. 14). Hence, kinship, community, common experience, symbolism and spiritual association are fundamental attributes of persons of Afrikan descent worldwide (Kambon, 1992). These precepts reflect the Tehutian principle of correspondence (the individual within the community).

Wilson (1999) agrees with these theorists. He sees culture as an instrument of power. His view was that persons of Afrikan descent will never solve their problems by integrating with their enemies. He concluded that it is a fantasy to believe that one day persons of Afrikan descent were going to merge with persons of European descent and become "invisible" (Wilson, 1999, p. 86). Wilson refers to invisibility in the sense that persons of Afrikan descent will never lose their racial, cultural, and spiritual distinctness and blend in with persons of European descent. (The concept of invisibility has also been used to explain the feeling, by persons of Afrikan descent, that they do not exist in the eyes of those of the oppressive race). Wilson (1999) states that Afrikan culture does not exist in reaction to the abuse and domination that European cultures have attempted to assert over peoples of Afrikan descent. Rather, Afrikan culture is designed to advance the interests of Afrikan people and operates in their best interests. In Tehutian terms, Wilson's concept of culture resonates with the principle of vibration, people of Afrikan descent vibrating in harmony with Afrikan culture, but not in harmony with European cultures as expressed through their political and social institutions. Olomenji (1996) agreed with Wilson, believing that the European system would destroy persons of Afrikan descent if the latter allowed them to do so.

Dixon (1976) states that "In the Euro-American worldview, there is a separation between the self and the nonself, or the phenomenal world. Through this process of separation, the phenomenal world becomes an object, an It'" (p. 55). Dixon (1976) goes on to note the contrast between the European and Afrikan worldviews. The European worldview is made up of values focused on "Doing, Future-time, Individualism, and Mastery-over-Nature" (p. 61) whereas the Afrikan worldview centers on man as part of nature experiencing, internalizing and personalizing the world: "The self is one with it" (Dixon, 1976, p. 61). Dixon's analysis of the Afrikan and European worldviews is based on differences in three areas of philosophical inquiry: axiology, the study of the values one holds; epistemology, the study of how one knows what they know; and logic, the study of how one organizes what they know (Dixon, 1976, p. 51). To this triad, King (1980) added the concept of ontology, the accounting for the essence of reality from an Afrikan perspective. Nichols (1974) used these concepts to create a matrix comparing persons of Afrikan descent to those of Native American, Latino/Latina American, Asian, Asian American, Polynesian, European, and European American descent. These comparisons resonate with the Tehutian principle of polarity. The European and Afrikan views are at opposite poles of the concept of what it means to be human. The difference is that the Afrikan view is more human and humane, while the European view is expressed in terms of power and control (Ani, 1994). Myers (1993) utilized these concepts to devise a schema of differences between western and Afrikan philosophical and conceptual systems. She added the concepts of process, identity, self-worth, values guiding behavior, sense of well-being, and understanding of the continuum of life to the Dixon/King/Nichols model.

Afrikan epistemology differs from the European version in that it does not rely solely on reasoning, but is rather uses a combination of feeling and thinking as a means to knowing. To explain this difference, Dixon coined the term "diunital," which means something that stands apart and is united with the other entity at the same time (Dixon, 1976). Diunital logic is characteristic of persons of Afrikan descent, whereas persons of European descent utilize an either/or logic that is characteristic of their worldview. Diunitality is an expression of the Tehutian principles of cause and effect and polarity, where the cause is united with the effect, and where the poles are at the same time opposing manifestations of the same truth. By contrast, the either/or logic of the Europeans separates the cause from the effect. The Afrikan orientation results in individuals finding their humanity as part of a social order, rather than in isolation from others (Dixon, 1976). In Dixon's view, the Afrikan worldview contains an orientation toward time in which past, present, and future are continuous, in contrast to the European worldview, which focuses primarily on the future (Dixon, 1976). In the frameworks evolved by Diop, Dixon, Myers, Nichols, and Nobles, one sees the Tehutian principle of polarity being expressed. In each of the analyses, the Afrikan and European worldviews are seen as opposites, and in being so, are two poles of larger truths, which are the possibilities and problems to be found within humanity. To date, due to the system of anti-Afrikan domination, the European pole appears to be in control globally. Scholars such as those mentioned above, have sought the opposite intellectual pole to enunciate a worldview that is human, humane, and centered in Afrikan reality.

Carruthers (1996) persuasively argued that the scientific method, a western invention derived from the European worldview, is not useful to persons of Afrikan descent in finding solutions for their challenges. Rather, he urged that the collective wisdom of people of Afrikan descent be gathered, and using that wisdom as the basis for research, to find and implement solutions. It is that call to which theorists of Afrikan psychology have responded. Afrikan humanists oppose Carruthers' perspective. They say that worldview is irrelevant when daily survival is the issue. Therefore, they conclude that persons of Afrikan descent do not have time for worldview discussions and thoughts (Allen, 2005; Ssekitooleko, 2005). The crises that exist in the Afrikan communities in the United States lend some credence to the view of the Afrikan humanists. However, without a framework from which to understand and analyze problems, people of Afrikan descent will find few lasting solutions.

The views of the radical school offer important vantage points from which to understand Afrikan psychology, for they summarize the pain, frustration, and disbelief that the subjugation of persons of Afrikan descent has wrought. In maintaining the humanity that forms a basis for the Afrikan worldview, persons of Afrikan descent are driven by their circumstances to try to make sense of the centuries and millennia of inhumane treatment that they have endured at the hands of those of European descent. It is their collective hope that the Tehutian principle of polarity operates with the principle of rhythm, and that the pendulum that has swung toward the negative will continue swinging, and will rise to a place that is positive for peoples of Afrikan descent.

Afrikan psychology: Theories

Theorists in the radical school have propounded varying psychology in concert with the main tenets of Afrocentric thought as described above. Afrikan psychology is based on a communal view of life, where importance is focused on the group, or the extended self, as enunciated by Nobles (1972). The individual finds their place, purpose, and fulfillment within the context of the group. These Afrikan-centered tenets are consistent across ancient, traditional and modern Afrikan societies (Grills, 2004a). Allen and Bagozzi (2001) said that persons of Afrikan descent form a more collectivist sense of self than persons of European descent. Amuleru-Marshall (1993) points out the connection in the Afrikan community between the individual, their family, and the "ethnocultural and sociostructural experience of the racial group" to which that person belongs (p. 24). Morris (1993) posits that for a people to maintain themselves as a group, they must supply themselves with sustenance, security, sex, and shelter (p. 70). Allen (2001) summarized DuBois's (1961) concept of a double consciousness by noting that DuBois found in persons of Afrikan descent an Afrikan cultural essence that sustains them individually and collectively when faced with attempts to dehumanize them. Wright (1984) provided a contrast to these Afrikan views on the extended self through his analysis of the European system of behavior, which he concluded was both individualistic and destructive. He stated that Europeans were psychopathic in their behavior toward persons of Afrikan descent, and that this psychopathology was transmitted biologically from generation to generation of persons of European descent. He coined the term "Menticide" to explain this pathology. "Menticide . . . is defined as the deliberate and systematic destruction of a group's minds with the ultimate objective being the extirpation of the group'" (Wright, 1984, p. 17).

Akbar (1976) also emphasizes the primacy of the group among people of Afrikan descent. He notes that persons of Afrikan descent heal each other and restore order as a group, whereas the West isolates the person to restore the self. Akbar (2004) discerned four mental disorders from which persons of Afrikan descent suffer. The first of these, alien-self disorder, results when a person of Afrikan descent has fully embraced the European worldview at the expense of an Afrikan worldview. In essence, they are disconnected from their culture and the Afrikan community. Focused on material acquisition, these people of Afrikan descent forego dealing with issues of concern for people of Afrikan descent, such as racism. Anti-self disorders result when a person of Afrikan descent is alienated from their Afrikanity and are hostile to anyone and anything that expresses an Afrikan worldview. This disorder can also be viewed as an extension or type of alien self disorder. Self-destructive disorders also reflect the negative impact of European domination on persons of Afrikan descent. Persons afflicted with this disorder engage in negative behavior of an individualistic nature, attempting to survive their oppressed situation by inflicting destruction on themselves or others. Organic disorders result when persons of Afrikan descent present psychological symptoms of societally created deprivation, such as malnutrition, exposure to hazardous chemicals, drugs or alcohol, or incarceration (Akbar, 2004). Akbar believes that persons of Afrikan descent operate in a communal rhythm. Therefore, when disorder occurs, whether the disorder is physical, mental, or spiritual, that rhythm has been disturbed and must be reestablished for healing to occur (Akbar, 1976). The centrality of rhythm in Akbar's theory resonates with the Tehutian principle of rhythm, where all operates in or near balance. When something occurs that moves the group out of balance, they act so as to bring the group back into balance. Akbar also notes that persons of Afrikan descent place value on submission to each other. This creates an equilibrium of opposites, or "twinness," such as man/nature or male/female (Akbar, 1976). This aspect of Akbar's theory resonates with the Tehutian principle of polarity, two seeming opposites that are part of a larger truth. In addition, the principle of gender is implicated, in that there are male and female and aspects to twinness, to Afrikan communities, and to the identities and personalities of persons of Afrikan descent.

Graham (1999) offers a view of Afrikan psychology via an analysis of the Afrikan worldview, which is seen as a system that is reinforced through aspects of culture such as "rituals, dance, storytelling, proverbs, metaphors, and the promoting of family, rites of passage, naming ceremonies, child rearing, birth, death, elderhood, and values of governance" (pp. 111-112). The values associated with this worldview include: "the interconnectedness of all things; the spiritual nature of human beings; collective/individual identity and the collective/inclusive nature of family structure; oneness of mind, body, and spirit; and the value of interpersonal relationships (Graham, 1999, p. 112). Within this worldview, the concept of oneness includes the unborn as well as all of those who have made their transition from this existence. In addition, the individual is not defined as a separate entity, but rather is defined in the context of their people. As Mbiti (1970) notes, "I am because we are; and since we are, therefore I am" (p. 141). Graham (1999) coined the term "twinlineal" to explain the integration of lineages from the mother and father that is characteristic of some Afrikan societies (p. 116). The point here is that children are the responsibility of the entire community, and not just of their parents. Fu-Kiau and Lukondo-Wamba (1988) concur, noting that the child does not exist outside the context of their community. Fu-Kiau and Lukondo-Wamba further state that the process of babysitting is a metaphor for the transfer of the values of the culture from one generation to the next. The centrality of spirituality in Afrikan psychology, and the connection between ancestors, those alive at present, and those yet to be born, all resonate with the Tehutian principle of correspondence: as above, so below. People of Afrikan descent focus on the interrelatedness of all, rather than the separateness of each thing and person, which is a hallmark of European thought.

Azibo (1988) proposed a three-part theory of the Afrikan personality. The inner core consists of those aspects of personality that are biogenetically determined and unchangeable. They consist of spirituality, energy, rhythm, and a self that is both personally defined and collectively determined by the larger community of persons of Afrikan descent (Azibo, 1988). His definition calls forth the Tehutian principles of rhythm, gender, and polarity. It echoes a group of theorists of Afrikan descent who focus on the centrality of the Afrikan essence or unconscious which is referred to as "the Black

Dot" (Bynum, 1999; King, 1994, p. 9). In Azibo's framework, the outer core consists of values, beliefs, attitudes and opinions that form a belief system and produce behavior in concert with the inner core when the external environment is unencumbered by unnatural and oppressive conditions (Azibo, 1988). The action component consists of behavior which grows out of and is influenced by the need for persons of Afrikan descent to survive and prosper (Azibo, 1988). This action component encompasses Asante's concept of Afrocentricity. The inner core aspect of Azibo's theory corresponds to the Tehutian principle that All is Mind/Spirit in the individual and communitarian sense. His outer core concept also resonates with this principle, as well as with the principle of rhythm. Azibo's component of action is part of the principle of vibration, which posits that everything moves or vibrates (acts) so as to reconcile apparent paradoxes. This also reflects the principle of polarity.

Kambon (1992) elaborated on the Afrocentric approach to psychology. He offered a four-part analysis of personality from an Afrocentric perspective. In his framework, the core of the Afrikan personality is biogenetic, a personality that has characteristics that are inherited from one's parents, who possess and reflect characteristics passed down from one's ancestors. As such, this personality cannot be altered by external influences, such as social or environmental forces. Kambon's framework assumes that personality has an intrapsychic aspect and an external aspect that are inseparable from each other or from the totality of the person. Further, personality is collective and social. As such, persons cannot be analyzed in isolation from those with whom they interact or from their community. Finally, personality can be defined by racial types based on the polarity of racial, biological, psychological, and behavioral traits. Kambon has yet to elaborate on his theory and to provide examples.

From this platform, Kambon proposed an "African Self-Extension Orientation" or ASEO, that is biogenetic, innate, unchangeable and deep-rooted energy that exists in all persons of Afrikan descent. This ASEO extends from the personal to the communal within the Afrikan context. Within the ASEO is a storehouse of Afrikan collective knowledge from all Afrikan ancestors that awaits use in attaining communal unity. This aspect of Kambon's theory sounds like aspects of Jung's theory of the "collective unconsciousness" (Elkins, 1998) springing from an Afrikan context. This ASEO is reflective of the Tehutian principle that All is Mind/Spirit. At the conscious level, Kambon proposes the existence of an "African Self-Consciousness" or ASC (Kambon, 1992, p. 54) that actuates and embodies the ASEO. ASC is a conscious expression of the core Afrikan spirituality, the ASEO. Kambon states that possession of the neurochemical melanin, which manifests in the darker skin of persons of Afrikan descent, is necessary for optimal human activity and for uniquely Afrikan expression and behavior. This analysis is problematic, as all members of the human family possess melanin of differing types and in differing degrees. Other than the obvious differences in skin tone between persons of Afrikan descent and those of European descent, many of Kambon's claims for the melanin found in persons of Afrikan descent have yet to be verified. Equally as important, research has not been done to determine what percentage of Afrikan ancestry must be possessed by persons designated by persons of mixed Afrikan and European heritage to receive full benefit from the presence of this Afrikan melanin. Ebonics, an Afrikan pattern of oral expression in the United States, contains a rhythm and symbolism that is uniquely Afrikan. These same characteristics are to be found in Afrikan music, dance, and body movement as forms of communication. These constants reflect the operation of the Tehutian principle of rhythm, the flowing of life, and the compensation for deviations from a central point that establishes the concept of balance, or Maat (Karenga, 2002).

An Afrikan person's ASC can be severed from their ASEO if that person experiences oppression in their daily lives. The psychopathological disorder that results from this circumstance is called "Psychological/ Cultural Misorientation" (Kambon, 1992, pp. 136-138). For persons of Afrikan descent, this is a primary psychopathology; all other forms of pathology in persons of Afrikan descent stem from it. Kambon proposed five criteria necessary for Afrikan life. It is implied that these five criteria are also part of the healing process for those who are psychologically misoriented. The processes are: re-education toward a Pan Afrikan, nationalist consciousness; elimination of names that are not Afrikan; re-education toward and practice of the Afrikan-centered astrophysical study of the universe and Afrikan history; removal of all Eurocentric symbols and artifacts from the home and other places where persons of Afrikan descent gather; and re-establishment of Afrikan rituals, symbols and structures (Kambon, 1992, pp. 179-182). Kambon's theory incorporates the Tehutian principles of rhythm, cause and effect (everything occurs according to a law), polarity, and vibration. These suggestions are difficult and expensive to implement, and eliminate all those persons of Afrikan descent who cannot or do not accomplish all five criteria from the Afrikan community. Kambon does not offer a process to those who wish to engage in these actions, and proposes no intermediary stages. In short, his criteria do not appear practical for the majority of persons in the Afrikan community in the United States. The financial outlay to accomplish some of these goals would stop many individuals desiring to change from doing so.

Goddard (1993) focused on defining the Afrikan family. He notes that the basis of the family is consubstantiation, the idea that the essence of all is of the same spirit and energy. This concept is a restatement of Dixon's concept of ontology, and resonates with the Tehutian principle that All is Mind/Spirit. Goddard notes that the functional family for persons of Afrikan descent in America is an extended family, one that relies on the larger family (relatives, neighbors, and friends) to frame itself. This same point is made by Billingsley (1968). This extended family is one that relies on communitarian concepts such as Maat, the Kemite concept that stands for "truth, justice, propriety, harmony, balance, reciprocity and order" (Karenga, 1993, p. 86), and spirituality for its existence. Children and adults alike belong to a community of shared values, goals and sensibilities. Nobles (1972, 1976) discussed and elaborated the concept of an extended self, in which elders, adults, and children are joined by ancestors and those yet to be born in a continuous community. Azibo (1996b) accepts and expands upon these concepts to include all those of Afrikan descent who have lived in the past, the ancestors, and all those who will live in the future, those yet to be born. In this way, the extended self becomes part of the Afrikan collective past and the Afrikan collective future. All of these concepts operate with a self that is part of a community. As such, they reflect the Tehutian principle of the interconnectedness of all. Allen (2001) notes that when persons of Afrikan descent take part in activities and rituals that they initiate, they achieve a greater sense of self and of the group, since they are part of the group. Combining this with Azibo's continuum, such activities that include ancestors and those yet to be born offer a continuity of experience and conception that centers the person in the full community of Afrikan descent- past, present, and future.

Nobles and Goddard (1993) offered a model of Afrikan-centered psychology and a theory of Afrikan culture based on the concept that all things in the universe are spirit and are interdependent and interconnected, echoing the Tehutian principle that All is Mind/Spirit. Nobles and Goddard (1993) note that the primal operation of the universe is "rhythmic and harmonious," functioning via similarity or difference on the same or differing metaphysical planes (p. 116). These thoughts evoke the Tehutian principles of rhythm, correspondence (as above, so below), polarity, and gender (everything has masculine and feminine principles). Nobles and Goddard (1993)proposed an eight-part theory of Afrikan culture that consists of the following elements: Consubstantiation-- all elements (humans, animals, inanimate objects) are of the same substance; Interdependence-- all elements in the universe are connected; Egalitarianism-- relations are harmonious and balanced; Collectivism-- codes of conduct based on the idea of group and/or collective survival/advancement; Transformation-- change is movement toward a higher level of functioning; Cooperation-- Things function based on mutual respect and viability; Humaneness-- behavior is governed by a sense of vitalism and viability; and Synergism-- the notion that the sum of complementary actions is greater than the total effort of individuals (Nobles & Goddard, 1993, pp. 117-120). The presence of the Tehutian principles of rhythm, polarity, cause and effect, gender, and All is Mind/Spirit in this formulation are clear. This model will be of service in framing a theory of Afrikan-centered identity development in Chapter IV.

Bynum (1999) states that the historical and cultural roots of all human consciousness is to be found in Afrika. The first expression of this collective unconscious is to be found in the civilization built in

Kemet, an unconscious that Europe and Eurocentric science repressed throughout the world. Bynum compares the collective and familial ethos of Afrika with the individual and egocentric ethos of Europe, and concludes, as did Diop, that the Afrikan and European worldviews are dichotomous on most levels of beingness. Bynum says that the Afrikan unconscious consists of rhythm, spirit, bioenergy, community consciousness, and intrapsychic and interpersonal principles (diunitality as noted by Dixon). He points out that in Afrikan traditions, five generations of ancestors and the unborn are considered to be part of the living and part of the continuity of consciousness (see, also, Rathele & Mutwa, 2008). He states that "The Hermetic tradition is the oldest known and still practiced tradition that amplifies and puts forth this worldview" (Bynum, 1999, p. 101), the worldview that centers the collective unconscious in Afrika. His interpretation is that the goal of these Hermetic/Tehutian principles was the transformation of consciousness. This stance can be applied to the field of psychology, where the healing of the human being or group of human beings, transforming them from dis-ease to health, is the goal.

Lee and Bailey (1998a) also note that persons of Afrikan descent in America place a high value on kinship, cooperation, mutual respect, commitment, and spirituality. They point out how those males of Afrikan descent who recognize the need for emotional support do not seek counseling in the mainstream mental health system. Rather, they seek it from relatives, friends, religious leaders, and in gathering places for male activities, such as barbershops, bars, and fraternal social organizations (Lee & Bailey, 1998a). Grills (2004c) suggests that Afrikan psychologists need to develop an understanding of "African metaphysics, the meaning of spirit in relation to human beingness" (p. 6). Metaphysics is an important aspect of most Afrikan spiritual systems (Mbiti, 1970), and is central to Tehutian philosophy, but is dismissed as irrelevant by European-based psychology.

Afrikan psychological theories have many commonalities that distinguish them from mainstream or European-centered theories. Kinship, the continuity and inclusion of ancestors and those yet to be born, the centrality of spirituality in all aspects of life, and the contextualization of the individual within the group are all core aspects (Oshodi, 2005). Also central are the concepts of an Afrikan unconscious shared by all persons of Afrikan descent, and the belief in Maat, or balance and justice as core values and goals (Beatty, 1997).

Afrikan Identity Formation

When, in 1957, Martin Luther King spoke of being "maladjusted to oppression," he gave definition to the emergent identity of a person of Afrikan descent who recognizes his or her humanity despite the actions of those who were oppressing them (King, 1957). Cabral (1973) extended this definition to include culture as central to identity. In his explanation of the role played by women and men from his country, Guinea Bissau, who were involved in the struggle to liberate their country from Portuguese colonialism, Cabral comments:

> But if one accepts that culture is a dynamic synthesis of the material and spiritual condition of the society and expresses relationship between both man and nature and between the different classes within a society, one can assert that identity is at the individual and the collective level and beyond the economic condition, the expression of culture. (Cabral, 1973, p. 64)

Helms (1993) defines membership in a racial group as a core aspect of identity development. She posits that this is true regardless of that individual's racial identification, due to society's emphasis on reward and punishment based on racial indicators. Lee and Bailey (1998b) note that adolescent identity formation includes selecting role models with whom to identify. These role models are, according to mainstream psychological theory, unavailable to males of Afrikan descent in America.

In 1971, at about the same time as White (1970) called for the formation of a Black psychology, Cross (Cross, 1979; Hall, Cross, & Freedle, 1972) authored a five-part model that described the process of racial consciousness formation, or, as he called it in his model, moving from "Negro" to "Black." The first stage is the pre-encounter stage, which is an initial stage that each person of Afrikan descent goes through, an initial identity that is anti-Afrikan and pro-European, taking the Eurocentric paradigm as their model. The second stage begins when the person of Afrikan descent encounters a person or an experience that disrupts their previous understanding of themselves and their world, opening them up to a new understanding. The third stage consists of two substages: the immersion and emersion substages. In the immersion substage, the person "discovers" their Afrikan identity. In this process, they divorce themselves from everyone and everything connected with European/White people and focus exclusively on persons of Afrikan descent. In the emersion substage, a critical analytical view of mainstream society and of Blackness replaces the view that all persons of European descent are racists and all persons of Afrikan descent are good. The next stage, the internalization stage, occurs as the newly emergent Afrikan adjusts to their new racial identity. In this stage, the person is engaged in what might be called an existential struggle to find and perfect their Blackness. Cross (1979) calls this struggle "Weusi Anxiety' or anxiety over Blackness (weusi is the Swahili word for black)" (p. 109). The fifth stage is the internalization-commitment stage. This stage involves the solidifying of this new Afrikan identity by the process of living it and becoming comfortable with their new self. Thomas (1971; see also Cross, 1980) developed a theory parallel to that of Cross, emphasizing the same process of identity evolution. Milliones made minor modifications to the Cross/Thomas model, combining the immersion/emersion and internalization stages (Cross, 1980). Parham added to Cross's theory, suggesting that one could move back through the stages once the cycle was completed, a concept he called "recycling," remain in a particular stage throughout life, which he called "stagnation," or progress from one stage to the next over their lifetime, called "stage-wise linear progression" (White & Parham, 1990, p. 53). Cross emphasized that his focus was on persons of Afrikan descent living in the United States. The challenge presented by the Cross and Thomas models is that they assume racial militancy to be a passing phase in the evolution of identity during the internalization and internalization-commitment stages. Cross (1991) perceives Afrocentrism as a perspective rather than a paradigm, thereby denying the centrality of an Afrikan worldview in analyzing identity development. In essence, these theories culminate with the person transcending race. Furthermore, these models lack a framework for identity formation from birth through maturity (Azibo, 1988). As Hall (1999) notes that there is no reason to assume that the factors that underlie racial group identity exist in children prior to their recognition of race as a social construct. In fact, as well as in practice, one finds many clients who are both deeply immersed in their Blackness, or "militant," and in need of assistance to return to a balanced ability to operate internally and in the world effectively.

Brookins (2004) operates from the belief, held by many, that identity is achieved in adolescence by means of a self-conscious search. His Adolescent Developmental Pathways Model suggests that rites of passage be undertaken to affirm the transition. Many traditional Afrikan societies saw identity as being shaped before and at birth, and continuing through life and on to ancestor status (Fu-Kiau, 1991). Phinney offered a three-part model of ethnic identity development, including an unexamined ethnic identity; a search for ethnic identity; and achieving ethnic identity (Gillem, Cohn & Throne, 2001, p. 184). This framework is insufficient for use in communities of Afrikan descent, because it fails to address issues specific to those communities. It does provide a contrast to the Cross and Thomas models, and the Parham additions, in that once one attains identity, that identity does not continue to shift or recycle. However, the model is too general to be applied successfully to the Afrikan community in the United States.

Statistics indicate that males of Afrikan descent are an endangered species in the United States (Madhubuti, 1990; Staples, 1982). The disparities between males of European descent and those of Afrikan descent with regard to rates of employment and incarceration as compared to college have been and are striking (Lee & Bailey, 1998a). The consequences on the psychological health of males of Afrikan descent, and on Afrikan communities, are powerful. It is in this context that Franklin (1992) raises the issue of "invisibility" with respect to males of Afrikan descent. He argues that they are denied a place of honor and opportunity in American society. This results in internalized rage, since the external expression of anger by males of Afrikan descent is not allowed or results in incarceration. This analysis perceives males of Afrikan descent as being passive in the face of oppression, and offers no corrective or solution to their dilemma.

Watts, et al. (1999) offered a five-part theory of what they called sociopolitical development. Their five stages include: the acritical stage, where a person is oblivious to social inequality; an adaptive stage, where the person perceives inequalities but does not confront them; a precritical stage where inequalities are recognized; a critical stage, where the person seeks knowledge about these inequalities and concludes that they are unfair and that change is necessary; and the liberation stage, where action and desire coalesce in the movement toward the elimination of oppression. Here the person becomes an active agent in the transformation of themselves and society. This theory assists in clarifying the nature of the Cross and Thomas theories. Those theories can be seen as more sociopolitical, or focused on environmental and other external phenomena, rather than on the development of the identity of the individual. Nonetheless, Watts, et al. (1999) fail to place persons of Afrikan descent at the center of the analysis, a task that awaited those who absorbed the essence of Asante's (1980) theory of Afrocentricity, and applied it to psychological issues.

Nobles (1976) posited that the existence of melanin in persons of Afrikan descent created an interdependence due to the complementarity of each such individual's sensory system. As a result, the individual self is part of and inseparable from the group, an extended self. Nobles' use of melanin as a correlate of heightened emotional arousal in persons of Afrikan descent (Clark, et al., 1976) avoids the controversy and confusion that Kambon's (1992) discussion inspires.

Azibo (1988) concludes that all of the theories of identity development created in the 1960s and 1970s are part of the negativist school, meaning that they do not take a positive perspective when viewing persons of Afrikan descent, and therefore are limited. He notes that these theories confine themselves to those who lived in the era of the Afrikan consciousness movements and attendant rebellions. He criticizes the end stages of these theories, such as those of Cross (1979) and Thomas (1971), where individuals are assumed to be integrationists who have moved beyond race because they have focused on their environment and their place in it, rather than on the will to survive which is the essence of persons of Afrikan descent. He ascribes these failures to an inability of these theorists to see the innate drive of Afrikan people to survive, called "survival thrust" by Baldwin/Kambon (Baldwin, 1986). This failure results in these theories reducing persons of Afrikan descent to reacting to an oppressive environment.

Semaj (1981) offered a theory on Afrikan extended identity, using the concept and terminology offered by Nobles. The first aspect of the theory is an alien extended self, which mirrors the alien-self described by Akbar. Such a person has lost their Afrikan identity and replaced it with an individualistic, materialistic identity based on a European worldview that denigrates them and their people. The second aspect is a diffused extended identity, which occurs when the person of Afrikan descent finds themselves believing in their Afrikanity but bowing to anti-Afrikan/anti-Black supremacy, believing that change is not possible. The third aspect is a collective extended identity which is centered on the survival and prosperity of persons of Afrikan descent (Semaj, 1981).

Williams (1981) used the Swahili term for Blackness, "WEUSI," as the framework for his three-part theory on personality. The first part is Blackness, which he designates as "WE," which refers to genetic, cultural, psychological and spiritual Blackness. These, in turn, comprise the basic attributes of all persons of Afrikan descent. Collectiveness, or "US" refers to the shared nature of Afrikan life on the individual, family, community, nation and worldwide levels. It involves working together toward common goals. Naturalness, or "I," concerns group traits that express themselves in individuals of Afrikan descent, such as spirituality and rhythm. The resonance with the Tehutian principles of rhythm, correspondence, and All is Mind/Spirit in this theory is strong. These are traits that empower persons of Afrikan descent. The centrality of spirit and spirituality in Afrikan-centered theories of identity development continues in all later theorists on Afrikan identity development. Webb-Msemaji (1996) believes that the spectrum of self-esteem, from low to high, may well be a measure of spirit. In so doing, he invokes the Tehutian principle of polarity-- everything has its opposites, which are identical in nature (self-esteem) but different in degree (low or high).

Azibo (1988) developed a nosology for use in rehabilitating those persons of Afrikan descent who find themselves in distress. Within the nosology, he calls self-destructive behaviors those lifestyles and behaviors that weaken the ethos of the Afrikan community, the survival thrust and the collective advancement of Afrikan peoples. Those who engage in such behavior are prone to experience anxiety over their collective identity, as they lose the cultural grounding in the Afrikan community that is the preventative for this condition (Azibo, 1988). Another misorientation occurs, according to Azibo, when persons of Afrikan descent believe in, and practice religions that deny Afrocentricity, Afrikan history, and Afrikan spiritual values, such as harmony with nature and denial of the spiritual essence of humanity (Azibo, 1988), the first Tehutian principle.

As part of the framework of Afrikan culture, Wilson (1999) notes that persons of Afrikan descent exist for the social unit and not for themselves. Kambon (1992) concurs, believing that all persons of Afrikan descent are innately connected with and drawn to the Afrikan personality. A person with a strong Afrikan personality, in Kambon's view, has a strong collective identity that consists of a collective sense of self and of Afrikan spirituality at their core. Akbar (1991) defines Afrikan male identity as consisting of self-respect, self-knowledge, and self-definition. He also states that when an Afrikan man declares his manhood, he stands in opposition to the European system that chooses to define him as less than a man. Kambon (1992) refers to this posture as forceful resistance to anti-Afrikanness in all forms, and retribution for the harming of Afrikan life (Kambon, 1992). These positions operate at two levels at the same time- that is, they are diunital. They frame an identity that is grounded in Afrikan concepts of what a man and a woman are based on Afrikan philosophical standards, and simultaneously reject the European framework that sees persons of Afrikan descent as socially and intellectually inferior. As such, it is to these concepts that one must turn to develop a coherent theory of Afrikan identity development.

Afrikan Psychology: Conclusion

According to the adherents of the Afrikan-centered psychological perspective, the current imbalance in the distribution of power cannot be addressed until Afrikan identity is defined, reclaimed, and strengthened; cultural assimilation and accommodation with American mainstream society are not viable options (Holdstock, 2000). Asante (1980) sees centeredness in the collective history of Afrikan people as a prerequisite to finding identity. To date, Afrikan male identity development has not been adequately delineated. There is no program, series of rituals or practices that can raise a boy of Afrikan descent to manhood. Such a theory requires development.

The issue of identity development in the Afrikan context has been the subject of many forms of Afrikan creative expression. Goddard (1993) makes note of the effectiveness of role models such as teachers and parents in promoting positive behavior on the part of youth of Afrikan descent. Nobles and Goddard (1993) extol the virtues of using proverbs as a teaching tool to connect a principle with a situation. They discuss the value of using a known commodity, such as the memory of a proverb, to understand a new or unknown situation as an important process. The I AM Model builds on that belief, positing that literature produced by persons of Afrikan descent can be a vehicle for creating understanding of a known (the situations in the book, poem, play, or proverb), and applying it to an unknown or unrecognized situation to assist a person of Afrikan descent who is in distress to find their way to wholeness and balance. Ellison's novel, *Invisible Man*, will be used here to demonstrate the viability of this process. Neal (1968) hinted at this possibility when, utilizing the double consciousness concept (being both Afrikan and American at the same time) enunciated decades earlier by W.E.B. DuBois, he noted that all writers of Afrikan descent in America have to take a stance either for it (being both Afrikan and American) or against it, either consciously or subconsciously, in their writing. In asserting that the writings of young Afrikan authors of the 1960s were intended to consolidate the Afrikan personality, he set the stage for the wedding of psychology and literature.

CHAPTER III
Afrikan American Literature and *Invisible Man*

No people on this Planet Earth, will ever be able to free themselves from their colonial past unless they embrace *their own* cultural heritage. This means that, they *must* have *their own* religious and political institutions, economic systems, including *their own* values of art and beauty. They would have to set values which are acceptable to themselves, whether they meet the values of others or not. (ben-Jochannan, 1963, p. 142, *italics in original*)

Introduction: Afrikan American Literature: Historical Background
Thrall, Hibbard and Holman (1960) place the origin of the novel in ancient Greece and Rome, while Beckson and Ganz (1989) focus more narrowly on the birth of the English novel, which they place in the early eighteenth century. Among the subcategories of novels, Thrall, Hibbard and Holman (1960) include the psychological novel, the political novel, the historical novel, and the novel of character. All of these forms are found in Ellison's novel, *Invisible Man*.

Prior to engaging Ellison's text, an overview of the history of Afrikan American literature is in order. Born of the trials and tribulations of the "maagamizi" or massive intentional destruction of enslavement (Karenga, 2002, p. 397), the first writings by persons of Afrikan descent in America were poems (Franklin & Moss, 2000). As Afrikans in America began to gain freedom from enslavement, either via manumission or escape, they expanded the forms of their written expression, writing narratives about their enslavement and their freedom (Gates, et al., 2004). So tenuous was the freedom of the escaped or fugitive slave author, that some wrote narratives in which they changed the names of the persons/characters in the narrative to avoid being found and re-enslaved.[11] The first novel known to be written by a person of Afrikan descent in America, Wilson's *Our Nig* (1859/1983), was published just prior to the outbreak of the Civil War in 1859. It was long lost to history as a novel written by a person of Afrikan descent, and has only recently been rediscovered and reissued.

As Afrikans in America were freed of the formal bonds of enslavement, and sought to expand their opportunities by gaining education, their opportunities to write creatively expanded. Publishers were initially hesitant to publish works by persons of Afrikan descent, seeing no audience and little profit in them (Robinson, 1971). However, eventually a market for their writing was realized, and their writing made it into print with increasing frequency (Bone, 1958). The attraction of and support by wealthy Americans, primarily of European descent, to the writings of persons of Afrikan descent in the early twentieth century led to a flowering of all forms of creative production (Kellner, 1979). This period was named after its artistic capitol, Harlem, New York, and is therefore commonly known as the "Harlem Renaissance" (Smith & Jones, 2000). Creative production by persons of Afrikan descent during this period centered on the defining, redefining, and refining of Afrikan cultures throughout the world, with a focus on Afrika (Karenga, 2002). Inspired by Marcus Garvey's Universal Negro Improvement Association (Martin, 1976) and interpreted by Alain Locke (1975), the Harlem Renaissance sprouted writers whose foci in their writings were the connectedness of Afrikans worldwide, describing and preserving Afrikan-based cultures, and racial pride (Kellogg, 1925).

Marcus Garvey's influence spread from Jamaica and the United States to the continent of Afrika, where his ideas, and those of the Harlem Renaissance writers, inspired Afrikan writers to begin the self-conscious creative search for their Afrikan selves and to put those ideas into words (Clarke & Garvey, 1974). Out of this sprang the Negritude movement, using a term coined by Cesaire and Senghor in the 1930s (Cesaire, 1969). Negritude is, in modern terms, the assertion by persons of Afrikan descent of the validity of their own humanity, culture, and contributions to civilization (Hawthorn, 1992). The cross-pollination of ideas and aspirations among scholars and authors of Afrikan descent from varying parts of the world was assisted by the periodic meeting of these intellectuals and writers in Pan-African Congresses, which began in 1900 in London (Padmore, 1972). The resolutions of the first Pan-African Congress called for the ceding of control of land, capital, labor, education, and government by Europeans to persons of Afrikan descent in areas of the world where persons of Afrikan descent were the majority (DuBois, 1946/1965). When the fifth Pan-African Congress was held in Manchester, England in 1945, Afrikan scholars and writers from around the world had rallied to the cry for recognition, equality and independence from colonization (DuBois, 1946/1965).

After World War II, few Afrikan writers were able to get their creative work published in the United States. Richard Wright was at the height of his literary power, having published *Uncle Tom's Children* (1936), *Black Boy* (1937), *Native Son* (1940), *12 Million Black Voices* (1941), and *American Hunger* (1944), all within a decade. Perhaps Wright's greatest contribution to writing by persons of Afrikan descent in America was his monograph, "Blueprint for Negro Writing" (1937/1978), which laid out a framework for writing by persons of

Afrikan descent. Wright believed that these writers should focus on culture, nationalism, and social consciousness in their writings. He saw a distinct nationalism in people of Afrikan descent, a nationalism that was reflected in the entirety of their culture, but especially in their folklore. Writers of Afrikan descent, he concluded, required a perspective in their writing that came from the struggles of their people. Further, these writers needed to work together in a collective effort for the advancement of the race, as opposed to writing creatively but in isolation from other writers of Afrikan descent (Wright, 1937/1978). Wright would depart from the United States for France in 1946, where he spent much of his time until his death in 1960. However, his influence over writers of Afrikan descent in the United States was enormous (Gates & Appiah, 1993). Wright's naturalist or realist literary style, as critics of European descent called it (Beckson & Ganz, 1989; Hawthorn, 1992), or "Urban Realism," as Karenga (2002) calls it, presented the stark physical, emotional, and psychological reality of fictional characters and their environment. Petry (1946, 1947, 1953) and Baldwin (1948) adopted and adapted this style, as did Ellison (1947). Wright's influence is also found in writers from the throughout the Afrikan world, such as Lamming (1954) from Barbados and Laye (1954) from Guinea (then called "French Guinea"), who adopted a similar style. In so doing, they were moving away from the prior generation of writers of Afrikan descent, whose style was less stark in its portrayal of the everyday life of persons of Afrikan descent (see Hurston, 1937, for example). In fact, Wright presaged the rallying cry of a future generation of persons of Afrikan descent with the publication of his book *Black Power* in 1954. This book on the search for independence by the country that became Ghana (then called the "Gold Coast"), called for nationalism by persons of Afrikan descent in all spheres of human activity (Wright, 1954). Here we can see the emergence of what was called a Black and Afrikan-centered approach to the creation of literature. These authors, led by Wright, and inspired by Garvey and DuBois, moved from using European forms of writing and European subjects, to Afrikan subjects and forms, in a conscious attempt to free themselves of the literary and psychological bonds that held previous generations of writers of Afrikan descent. This racially nationalistic stance was assumed by these authors as artists, and as women and men. They used their art to uplift themselves and their people, guiding and often leading the movement of people of Afrikan descent for freedom and self-determination.

In the field of poetry, Gwendolyn Brooks published *A Street in Bronzeville* in 1945, which Wright praised for its realistic portrayal of persons of Afrikan descent (Melhem, 1987). One finds a similar evolution in music by persons of Afrikan descent in America, where the musical forms of the blues and jazz evolved into more politically and socially conscious art forms (Jones, 1963). It is in this era, and in this milieu, this beginning fusion of forms to respond to the dictates of conscience and consciousness by persons of Afrikan descent, that, in 1947, Ralph Ellison's monumental novel, *Invisible Man*, was produced (it was first published in 1952). Though Ellison's personal vision and ideological position differed markedly from that of Wright, he is still considered part of the Wright "school" of writers of Afrikan descent (Wall, 1994).

This discussion will use Ellison's *Invisible Man* as a metaphor and example of how to construct a theory of identity development using a literary character to inform psychological theory. *Invisible Man*'s main character is engaged in a journey toward self-consciousness and self-awareness, both personally and racially. In this chapter, his quest will be explained, discussed, and analyzed, blending in Afrikan-centered psychological theories. In the next chapter, an Afrikan male identity development framework, the I AM Model, will be constructed and analyzed using this character. The usefulness of the framework in analyzing issues of identity development in Afrikan American males will be discussed, and suggestions will be offered for using this framework in therapy with males of Afrikan descent.

Invisible Man: The Novel

For purposes of this project, *Invisible Man* will be discussed out of order. It is written in the form of an extended flashback. That is, the first and last chapters of the novel, entitled "Prologue" and "Epilogue," respectively, take place in the present. The remainder of the book, chapters 1 through 25, takes place in the past, and explains what is occurring in the prologue and epilogue. The novel will be discussed chronologically to clarify how the issue of identity development is exemplified. Thus, chapters 1 through 25 will be discussed first, and then the prologue and epilogue will be discussed.

The first, and clearly the most striking, aspect of *Invisible Man* is that its main character has no name. In Afrikan cultures, words link the parts of the community together, and link the human community with God (Carruthers, 1995). Traditionally, a newborn was named after a desired quality or characteristic, or after an aspect of their emergent personality (Karenga, 1975; Onyefulu, 2004). Afrikan societies used differing ways to name children. Among the Akan of West Afrika, children were named according to the day on which they were born, known as the giving of "day names" (Thornton, 1993, p. 727). Thornton (1993) notes that Kongo parents gave their children double names to indicate from whom they were descended. Angolan parents used both family and descent names (Thornton, 1993). Holloway (2006) notes that in Afrika, real names given at birth were to be kept secret, lest they be used to work bad magic on the person so named. He also notes that an Afrikan person's name could change over time, especially when the person passed through to a new stage of life, and that nicknames, a common feature in families of Afrikan descent in America, reflect the Afrikan practice of providing a given name to be used in public, and a personal name for use within the family. Fu-Kiau (1991) notes that many Afrikan children have three names: an informal name, a formal name, and an initiation name. Texts have been published with Afrikan names and the attributes they describe so that people of Afrikan descent can give Afrikan names to their children.[12] In his book *Roots: The Saga of an American Family*, Haley (1976) illustrated how the Mandinkas of West Africa named their male children in line with the promise the child held for the future of the nation. It was believed that a child would develop up to "seven characteristics of whomever or whatever he was named for" (Haley, 1976, p. 2). Ssensalo (1978) points out the crucial place that naming had during the holocaust of enslavement. Slave masters denied the use of Afrikan names to Afrikans as part of a systematic attempt to strip Afrikans of their culture and traditions. In their place, these Afrikans were given European first names, and denied a last name, which, in the United States, denotes family history and respectability (Ssensalo, 1978). When these Afrikans gained their freedom, they gave themselves new names, indicative of their new status: freed from bondage. Elijah Muhammad, leader of the Nation of Islam, proposed that people of Afrikan descent in America discard the names given to them by their former slave masters, for "A good name is, indeed, better than gold" (Muhammad, 1965, p. 55).

Hence, in the Afrikan community, names are of symbolic as

well as familial importance. The main character in *Invisible Man* <u>does</u> have a name. His name is discussed but never revealed in the course of the novel. As a result, critics and commentators have chosen a variety of substitutes in an effort to name him. If one is invisible, one must also be nameless, for to be named implies that one exists. Baldwin took Ellison's lead in two volumes of social commentary, which he entitled *Nobody Knows My Name* (1954) and *No Name in the Street* (1972). Since the book, *Invisible Man*, exists, as does its main character, a means had to be developed by critics to discuss this character. As the novel is conveyed in the first person and from the perspective of the main character, many choose to call him the narrator (Gibson, 1971; Walling, 1973). Others refer to him as "P" or the protagonist (Gray, 1978). Still others refer to him as IM, short for (I)nvisible (M)an (Wolfenstein, 2003). This latter perspective is symbolically important, as the seeking of identity is the process of figuring out who one is, or, in the first person, who "I am." It can also be seen as the contraction "I'm," which implies possession of something. The "something" in this case, is his own identity. Hence, IM is appropriate for one who is invisible- that is, one who is not, but who is trying to become one who is. "I am" answers Fanon's "In reality, who am I?" (1963, p. 251), the question that one subjugated by oppression asks. In the 1960s, this phrase "I am" became popularized in the phrase "I Am Somebody," which the Reverend Jesse Jackson popularized in a poem by the same name in 1966 (Civil Rights Quotes, 2005). The phrase evolved, and in the 1968 strike of garbage collectors in Memphis, Tennessee (during which Dr. Martin Luther King was assassinated), it became "I *am* a man" (Bennett, 1971, p. 92). It appears that Ellison was aware of the "IM, I'M, and I AM" possibilities, because a friend of Ellison mentioned them in a letter to Ellison (Murray & Callahan, 2000, p. 32). Therefore, in referring to the main character, the term "IM" will be used.

Invisible Man: Chapters 1 through 25. Chapter 1 takes the reader back to IM's high school days. He is about twenty years old at the time of the prologue, the "present" of the novel. But in Chapter 1 IM has just graduated from high school, meaning that he was still an adolescent. Looking back, he says of himself: "I was naive. I was looking for myself and asking everyone except myself questions which I, and only I, could answer" (Ellison, 1947, p. 15). He goes on to say that he used to be ashamed of being the descendent of slaves, and then was ashamed of being ashamed. He recalls his grandfather, who had been a slave. On his deathbed, his grandfather says something to him that haunts him throughout the novel:

> Son, after I'm gone I want you to keep up the good fight. I never told you, but our life is a war and I have been a traitor all my born days, a spy in the enemy's country ever since I give up my gun back in the Reconstruction. Live with your head in the lion's mouth. I want you to overcome em with yeses, undermine em with grins, agree em to death and destruction, let em swoller you till they vomit or bust wide open.' (Ellison, 1947, p. 16)

This was an elder passing the knowledge of the group down to the new generation. As such, it is a beginning statement of the continuity between the past and the present. His grandfather's words are a prescription for survival and prosperity in a hostile environment. IM is puzzled by what his grandfather meant. IM mused that whenever things were going well for him, he felt guilty, as though he was carrying out this advice from his grandfather, which he still did not understand.

When he was praised by the White men of the town as being an example of proper conduct by people of his race, he could not figure out what treachery he was committing, as his grandfather, who was also praised, had said. IM gave a speech at his all-Black high school graduation in which he extolled the virtues of humility as the key to progress. In an aside, he notes "Not that I believe this-- how could I, remembering my grandfather?-- I only believe that it worked" (Ellison, 1947, p. 17). Here he is acknowledging that he subconsciously understands his grandfather's point, and is acting on it. He was invited to give the speech again to a gathering of the leading White citizens of the town. When he arrived, he found himself in a group of ten young men of Afrikan descent who were forced to look at a beautiful, naked White woman who had an American flag tattooed on her belly. The young men were then ordered to participate in a battle royal, where each of the young men of Afrikan descent were dressed in boxing shorts and gloves, blindfolded, and asked to fight the others in a ring for the amusement of the audience. IM's concerns were not with being hurt, but rather that such a display might detract from his speech. He noted that he fancied himself as the next Booker T. Washington.[13] IM notes that-- "Blindfolded, I could no longer control my emotions. I had no dignity" (Ellison, 1947, p. 22). In other words, he could no longer see, literally and symbolically. He was, at that moment, not visible to himself. Eventually he is forced to fight the biggest of the young men, symbolically pitting the most athletic against the most intellectual, with the promise of money for the winner. The crowd cheered as IM was beaten down. Then all the young men were invited by the White man to fight again, this time for coins that had been laid on a carpet. The carpet was electrified, however, and each young man was shocked as they touched the carpet in the attempt to retrieve the coins. Once this escapade was done, and all the White men were drunk, IM was allowed to give his speech. Beaten and swallowing blood as he speaks, he delivers his speech, quoting and paraphrasing Booker T. Washington's famous "Atlanta Cotton Exposition Address" (Washington, 1906). While so doing, he makes a mistake, speaking his mind instead of delivering his speech (divine speech rather than politically correct speech), and advocates social "equality" rather than "social "responsibility" for members of his race (Ellison, 1947, p. 31). He is quickly challenged by members of his white audience, and apologizes for the mistake. For his efforts and the position he took in his speech, IM is given a briefcase in which he finds a document giving him "a scholarship to the state college for Negroes" (Ellison, 1947, p. 32). Exhausted from his trials, he goes home and falls asleep. He dreams that he is with his grandfather at a circus. His grandfather asks him to open the briefcase and read what is inside. He finds a number of envelopes, each containing other envelopes. In the last one is a note that says: "To Whom It May Concern . . . Keep This Nigger-Boy Running'" (Ellison, 1947, p. 33). He awakes, not knowing what the dream, the note, or his grandfather's laughter in the dream meant. Dreams, which are an important part of the metaphysical world of Afrikan people, are the vehicle by which the ancestors communicate with those in the present (Mutwa, 1996). This and the other dream sequences in the novel illustrate Afrikan spirituality and the first Tehutian principle in operation, that change is the only permanent reality.

Chapter 2 takes the reader and IM to the Black college. IM gazes at a statue of the founder of the college, himself a person of Afrikan descent, appearing to lift a veil from a kneeling slave. IM muses: ". . . I am puzzled, unable to decide if the veil is really being lifted, or lowered more firmly in place; whether I am witnessing a revelation or a more efficient blinding" (Ellison, 1947, p. 36). This is an allusion to the question as to whether mainstream education in the United States frees the person of Afrikan descent, or binds them to service to others (Woodson, 1933). It also reminds the reader of the saying by Cheikh Hamidou Kane that opens Chapter I of this book: "Is what one learns worth what one forgets?" (Kane, 1969, p. 31). As he remembers the annual visit of the mostly white trustees to the college, IM remarks that those memories are part of a life that is now dead, and then he thinks, parenthetically "(Time was as I was, but neither that time nor that I' are any more.)" (Ellison, 1947, p. 37). In other words, this is a flash forward to the present. IM is noting that he is not now (in the present from which he is recalling these events) who he was then, when he was at the college. He does recall one of the white trustees who came each year to donate money for the survival of the college, Mr. Norton. It was IM's task to drive Mr. Norton around the campus when Norton visited the campus during IM's junior year. As Norton discussed Ralph Waldo Emerson and his ideas on self-reliance[14] with IM, he took a wrong turn, and ended up at the shanty of a local sharecropping family, the Truebloods. As they approach the home, IM recalls the hatred that the college students had for those who, like Jim Trueblood, the husband of the family, came to the college to sing periodically: "How all of us at the college hated the black-belt people, the peasants,' during those days! We were trying to lift them up and they, like Trueblood, did everything it seemed to pull us down" (Ellison, 1947, p. 47). Trueblood, by name, is an authentic person of Afrikan descent, a counterpoint to the educated persons of Afrikan descent, represented by IM. Here we see the Tehutian principle of polarity in operation, Trueblood and IM representing two poles of people of Afrikan descent at the college. IM absentmindedly tells Norton that Trueblood has impregnated both his wife and his daughter. Norton is enthralled by this revelation, and asks Trueblood to tell him how this occurred, and why he feels no remorse about it. Meanwhile, IM is disgusted with Trueblood's innocent interpretation of this event, and with himself for having brought Norton to this place. Trueblood says that once the people at the college learned of his transgression, they tried to make him and his family move. Trueblood continues:

> I went to see the white folks then and they gave me help. That's what I don't understand. I done the worse thing a man could ever do in his family and instead of chasin' me out of the country, they gimme more help than they ever give any other colored man, no matter how good a nigguh he was. (Ellison, 1947, p. 67)

As they leave, Norton gives Trueblood a one hundred dollar bill. Then, Norton tells IM that he needs a "stimulant" to recover from what he just learned (Ellison, 1947, p. 70).

Chapter 3 takes place at a bar/restaurant/whorehouse called the Golden Day, which catered exclusively to clients of Afrikan descent. IM drives Norton there to get him a "stimulant." While the college had tried to make the Golden Day respectable, the local White population was invested in it staying the way it was. When Norton and IM arrived, Norton was nearly unconscious, having been overcome with emotion upon hearing Trueblood's story of incest. The bar was filled with military veterans of Afrikan descent from the nearby insane asylum. The man in charge of the vets was named Supercargo, a name which means both the officer on a merchant ship in charge of the cargo (Merriam-Webster's Collegiate Dictionary, 2001), and which implies "superego" in the Freudian sense, here seen as restraining the vets from losing control of their egos due to the presence of alcohol and women (Merriam-Webster's Collegiate Dictionary, 2001). Many of these veterans had been professional people: doctors, lawyers, and teachers before being hospitalized. In this chapter, we see the operation of what Akbar called the alien-self disorder. The vets are focused on the acquisition of alcohol and women, rather than on acquiring a new consciousness. The bartender refuses to give IM a drink to take to Norton, who was in the car. So, with the assistance of some of the vets, Norton is carried into the bar. Once Norton has a drink and is somewhat revived, he asks if all of the men are patients of the asylum. One responds: "We're patients sent here as therapy . . . . But, they send along an attendant, a kind of censor, to see that the therapy fails" (Ellison, 1947, p. 81). In other words, instead of curing their insanity, trips to the bar are designed to reinforce it, as the construction of alien-self disorder would predict. Still stunned by Trueblood's story, Norton is laid down on the bed of one of the prostitutes to rest, and is attended to by one of the patients who said that he had been a physician. IM is fearful of trouble because this vet is talking to Norton freely and as an equal. In a society where those of European descent hold virtually absolute power, as was the case in the South in the early decades of the twentieth century, for a man of Afrikan descent to speak freely to a powerful White man was a certain sign of either insanity, or a fully recognized Afrikan humanity. The vet says to Norton: "To some (of the patients in the bar) you are the great white father, to others the lyncher of souls, but for all, you are confusion come even into the Golden Day" (Ellison, 1947, p. 93). This vet says that he had learned surgery in the war, but when he returned to the United States, he was beaten and driven out of town for saving someone's life. The vet then asks IM if IM understood what he meant by saying that Norton was both a savior and lyncher. IM does not understand. The vet turns to Norton and analyzes IM. He says:

He registers with his senses but short-circuits his brain. Nothing has meaning. He takes it in but he doesn't digest it. Already he is- well, bless my soul! Behold! a walking zombie! Already he's learned to repress not only his emotions but his humanity. He's invisible, a walking personification of the Negative, the most perfect achievement of your dreams, sir! The mechanical man! (Ellison, 1947, p. 94)

In other words, the vet, who is supposed to be insane, sees that IM, by accepting subservience to people of European descent such as Mr. Norton, is himself insane. The vet thus predicts IM's invisibility. (Note that the vet, too, lacks a name.) The vet concludes that IM "believes the great false wisdom taught slaves and pragmatists alike, that white is right. I can tell you *his* destiny. He'll do your bidding, and for that his blindness is his chief asset" (Ellison, 1947, p. 95). This blindness recalls the Founder's statue with the veil placed over the head of the slave. Norton and IM quickly escape the Golden Day, and head back to the campus.

Chapter 4 is full of foreboding for IM. As he drives Mr. Norton back to the campus, he thinks to himself: "Here within this quiet greenness I possessed the only identity I had ever known, and I was losing it" (Ellison, 1947, p. 99). For IM, being a college student as a lifelong goal, one that defined who he was, and these events with Norton threatened to end that existence. When they got back to the campus, IM made a halfhearted apology to Mr. Norton. As he went to get Dr. Bledsoe, the president of the college, IM expressed his desire to be like Bledsoe: "Influential with wealthy men all over the country; consulted in matters concerning the race; a leader of his people; the possessor of not one, but *two* Cadillacs, a good salary and a soft, good-looking and creamy-complexioned wife" (Ellison, 1947, p. 101). Norton absolved IM of responsibility for his discomfort, but IM knew that Bledsoe would not absolve him.

Chapter 5 takes place largely in the campus chapel, where IM must attend a special evening service held for the trustees. As he walks silently with other students toward the chapel, he cannot help but think they were going to a place of acceptance, where their allegiance was unquestioned. IM describes the love that accompanied that loyalty as: "Loved as the defeated come to love the symbols of their conquerors" (Ellison, 1947, p. 111). In a sense, the students were going to the chapel as much to worship at the altar of White power, as represented by the trustees, as they were going to worship on the spiritual level. IM loses himself in thought as he watches the guests on the stage. He compares himself to Dr. Bledsoe, the president of the college, as a leader of people of Afrikan descent, and as a person "who could touch a white man with impunity" (Ellison, 1947, pp. 114-115), a thought which, when ascribed to the vet, explained why the vet was insane. Bledsoe assumed a stature of mythic proportions at the college. He had arrived at the college destitute and had worked his way up ("bled so") from feeding the pigs as a student to the presidency of the college. As IM again pondered his own fate, he imagined "the return home and the rebukes of my parents" (Ellison, 1947, p. 117). A eulogy about the college's Founder is delivered by Reverend Homer Barbee, a blind man of Afrikan descent. The Founder's tirelessness and influence in building the college is reminiscent of the story of Booker T. Washington building Tuskegee Institute, the historically Black college in Alabama (Washington, 1906). Barbee describes how the Founder was asked by a crowd what they needed to do to alter their oppressed situation. As he begins to answer, the Founder falls, and died soon after, symbolically unable to provide an answer. Bledsoe took over the work of the Founder, keeping the dream of the college going, and building on the dream. When Barbee finished his sermon, Dr. Bledsoe led the students in a heartfelt hymn. Amidst the singing, IM thought: "I could not look at Dr. Bledsoe now, because old Barbee had made me both feel my guilt and accept it. For although I had not intended it, any act that endangered the continuity of the dream was an act of treason" (Ellison, 1947, p. 134). Expecting to be expelled, IM wonders "Where would I go, what would I do? How could I ever return home?" (Ellison, 1947, p. 135). IM is afraid to disappoint his parents, to the point that he would consider not going home and confronting them with his mistake. He is, in Tehutian terms, swinging toward the negative pole or is out of rhythm, out of sync.

In chapter 6, IM's foreboding is brought to fruition. He meets with Bledsoe in Bledsoe's office, and tries to explain his actions. Bledsoe says he had expected IM to lie when Mr. Norton ordered him to find "stimulant" for him: "*Please* him? And here you are a junior in college! Why, the dumbest black bastard in the cotton patch knows that the only way to please a white man is to tell him a lie!" (Ellison, 1947, p. 139). Bledsoe accuses IM of having brought the race down (in the eyes of Whites). IM threatens to tell Norton that Bledsoe was dismissing him from the college. Bledsoe reminds IM that while the trustees support the college, he controls it. He says to IM: "You're nobody, son. You don't exist-- can't you see that. The white folk tell everybody what to think-- except men like me. I tell them; that's my life, telling white folk how to think about the things I know about" (Ellison, 1947, p. 143). In other words, Bledsoe has taken on a role similar to that of the vet in the Golden Day. Speaking equally to the White trustees indicates that Bledsoe is as insane as the vet was supposed to be, or as full of guile. Bledsoe suggests that IM go to New York, work and earn some money, implying that IM could return to the college in the fall. Bledsoe offers to give him letters of introduction to donors to the school so that IM could get work. These suggestions are made by Bledsoe so that IM will not go home and tell his parents what has been done, and will not go to Norton and tell him what has transpired either. For if IM had done either of these things, Bledsoe would have been in trouble, having violated his responsibility toward IM, one of his prized students. As IM returned to his room to pack and to leave the only life he had known, he maintained his innocence in his own mind. Nonetheless, he saw two choices: accept responsibility for what had occurred, or face the world of Trueblood and the Golden Day. He leaves the next morning.

In chapter 7 IM begins a new life. When he gets on the bus heading North, he meets the vet from the Golden Day. The vet advised IM "...to look beneath the surface . . . . Play the game, but don't believe in it-- that much you owe yourself . . . . Learn how it operates, learn how you operate" (Ellison, 1947, p. 153-154). The vet goes on to say: "Be your own father, young man. And remember, the world is possibility if only you'll discover it" (Ellison, 1947 p. 146). The vet is telling IM to find his own consciousness, and to be his own family, because IM is leaving both his biological and adopted (college) family behind. The vet is operating from the perspective that kinship and belonging, Afrikan traits, are important, vital. The vet leaves the bus, and IM rides on to New York alone. When he arrives in Harlem, IM is stunned to see so many people of Afrikan descent. He encounters a speaker at a street corner gathering, who is threatening to drive White people out of Harlem. Noting that he had never seen so many men of

Afrikan descent angry in public, he finds his way to the Men's House (a place like the housing found at Young Men's Christian Association hotels), looking for a room to rent.

In chapter 8, IM goes to the offices of each of the persons to whom Dr. Bledsoe has addressed a letter of introduction in hopes of getting a job. Nothing comes of the first six attempts. IM had written home to tell his family that he had found work with a member of the college's board of trustees, thus lying to his family, who did not know that he had been expelled from the college. In so doing, IM had denied the counsel, warmth, protection, and connection that family provided him as a person of Afrikan descent. He continued to be out of rhythm. As the days wore on, no job had materialized, and his funds ran low, he began to doubt himself: "I grew conscious that I was afraid; more afraid here in my room than I had ever been in the South" (Ellison, 1947, p. 170). This fear grew out of his aloneness. He had cut off his connection with his family and his community, and was suffering the negative emotional consequences of this bad choice.

In chapter 9, IM goes to meet with the seventh person to whom Dr. Bledsoe had written a letter, Mr. Emerson. On the way, IM meets a man named Peter Wheatstraw, who is singing the blues as he pushes a cart filled with used blueprints. Wheatstraw engages IM in a delightful conversation, but IM understands little of it. When Wheatstraw says "Damn if I'm-a let em run *me* into my grave" (Ellison, 1947, p. 175), the reader is reminded of IM's grandfather's imprecation in the dream: "Keep This Nigger-Boy Running." When IM arrives at Mr. Emerson's office, he meets Emerson's son, who appears to be sexually attracted to IM. The younger Emerson is reading the book *Totem and Taboo*, Freud's treatise on the psychological challenges faced by primitive peoples (implying, here, persons of Afrikan descent) in comparison to those faced by civilized people (implying, here, persons such as Emerson); (Freud, 1946). Emerson asks IM about his career goals. IM responds that he would like to return to the college as a teacher and work his way up to being Dr. Bledsoe's assistant, following the path Bledsoe had earlier taken. As young Emerson attempts to take IM into his confidence, IM recalls his grandfather telling him: *"Don't let no white man tell you his business, cause after he tells you he's liable to get shame he tole it to you and then he'll hate you. Fact is, he was hating you all the time . . . . "* (Ellison, 1947, p. 186, [*italics in original*]). Again, IM is visited by his ancestor, his spirit guide, who is giving him information from the past that is vital to IM's present. But because IM is partially disconnected from his lineage and cultural heritage at the time, he gets the message, but does not understand its meaning. IM is concerned that the younger Emerson is stopping him from meeting the elder Emerson and prove that he is the person who Bledsoe has recommended in the letter. IM says that he will prove his identity, that he is the person referred to in the letter. Young Emerson responds: "Identity! My God! Who has any identity anymore anyway?'" (Ellison, 1947, p. 187). For persons of European descent, such as Emerson, power may have replaced identity. But for IM, identity is vital, and his search for identity is his central quest. His difficulty is that his identity is bound up in the identity of his people, and he has consciously, forcefully been separated from his people. Therefore, he is confused about his identity at this point. Young Emerson advises IM not to return to the college if he wants to do what is best for himself. IM responds: "But I know what's best for me . . . . Or at least Dr. Bledsoe does . . . .'" (Ellison, 1947, p. 189). Young Emerson eventually gives Bledsoe's letter to IM to read. The letter says that IM has been permanently expelled, and that he should not be told of this fact, and not be given a job. IM was stunned. Young Emerson offered IM employment as his valet, and told him about a possible job at the Liberty Paints factory. IM declined Emerson's assistance, but followed up on the lead about the Liberty Paints job, and got an interview without Emerson's assistance.

Chapter 10 finds IM at the Liberty Paints factory. Using Emerson's name without his knowledge, IM gets a job. His initial task is to work under Kimbro, a man described by one of the other employees as a "slave driver" (Ellison, 1947, p. 198). IM's task is to take a black substance and drop it into white paint to make Liberty's famous "Optic White" paint, which is used to paint government monuments all over the country. This paint formula is an allusion to race and the mixing of the races that is underplayed in America. It is also a metaphor for people of Afrikan descent, who built the wealth of the United States, but gained no recognition for it (the Black disappears in the white paint). Liberty employees are proud of their paint: "White! It's the purest white that can be found. Nobody makes a paint any whiter" (Ellison, 1947, p. 202). IM makes a mistake that ruins a small batch of paint (symbolically, a person of Afrikan descent cannot make anything pure white). He thinks to himself: "It was not all my fault and I didn't want the blame . . ." (Ellison, 1947, p. 202), echoing his feelings earlier concerning the incident with Norton at the college. Kimbro sends him back to the office be reassigned or released. IM was reassigned to work with Lucius Brockway[15], an old man of Afrikan descent, in a basement of the factory. Brockway is suspicious of IM from the start, believing that IM was after his job. He asks IM his name. "I told him, shouting it in the roar of the furnaces" (Ellison, 1947, p. 208). The roar of the furnaces prevents reader from "hearing" his name as well. IM's job is to watch the gauges on each of the furnaces to insure that they do not overheat. The product of these furnaces is the base from which all the paints were made. Brockway summarizes his job in this way: *"we the machines inside the machine'"* (Ellison, 1947, p. 217, [*italics in original*]). Brockway brags that he invented the slogan for the optic white paint: "If It's Optic White, It's the Right White'" (Ellison, 1947, p. 217), which causes IM to think "If you're white, you're right'" (Ellison, 1947, p. 218), a cliche for racial superiority in the United States. IM enters a union meeting which is in progress in the locker room where he goes to get his lunch. The union members, who refer to each other as "Brother," treat IM as anything but a brother, believing him to be against the union, partially because he worked with Brockway, who was against unions. As they finally allow him to go, IM thinks to himself: "I felt that every man looked upon me with hostility; and though I had lived with hostility all my life, now for the first time it seemed to reach me, as though I had expected more of these men than of others . . . ." (Ellison, 1947, p. 223). IM continues to be out of rhythm. He believes that unions equalized relations between workers regardless of race, because of their use of the term "Brother" when speaking among themselves. He learns that they were no different than other White men, as he receives the same hostility that he has encountered from people of European descent his entire life. When Brockway learns that he had been at a union meeting, he orders IM to leave and threatens to kill IM. IM recalls his childhood training, which was to respect the elders of his race, regardless of how foolishly they spoke or acted. This is in keeping with Afrikan cultural philosophy. But IM continues to be out of rhythm with his identity as a person of Afrikan descent, and quickly loses respect for Brockway. Brockway attacks IM, and IM knocks him down. As they joust, one of the furnaces overheats. Brockway instructs IM to turn the valve wheel the wrong way, and while Brockway escapes, IM is still in the area of the furnaces when it explodes. The reader is left with the implication that Brockway will blame IM for the explosion, thereby securing his job, and thwarting any future attempts to provide him with assistance.

IM appears to be an observer as he finds himself in the factory hospital in chapter 11. As the medical staff assesses him, he is unable to verbally respond to them. As he lay in the hospital, he recalled incidents from his youth, including a day that he saw " . . . the hounds chasing black men in stripes and chains" (Ellison, 1947, p. 234), an allusion to prisoners escaping from a chain gang in the South, seeking their freedom, as IM was. As IM undergoes electroshock therapy, one of the doctors says: "...I believe it is a mistake to assume that solutions-- cures, that is-- that apply in, uh . . . primitive instances, are, uh . . . equally effective when more advanced conditions are in question. Suppose it were a New Englander with a Harvard background?'" (Ellison, 1947, p. 236). This implies that persons of Afrikan descent were primitive, less valuable, and therefore more chances can be taken with their lives when treating them, such as using electroshock on him. The electroshock therapy machine was designed to simulate "a prefrontal lobotomy without the negative effects of the knife" (Ellison, 1947, p. 236), thereby altering the personality and making the patient docile. As an experiment, the doctors decided to apply extra electrical current to IM. The doctors then asked him his name, which he no longer knew. He also could not remember his mother. However, he was able to recall a childhood folktale, Buckeye or Brer Rabbit, when the doctors suggested it. That irked him: ". . . it was too ridiculous- and somehow too dangerous. It was annoying that he had hit upon an old identity . . . ." (Ellison, 1947, p. 242). That identity was one of a person of Afrikan descent outsmarting the master during the era of enslavement, the theme of the Brer Rabbit tales (Hamilton, 1985). Hence, while he had lost his connection to his family, and had lost his memory, he had not lost the core of his Afrikan consciousness, as represented by the Buckeye (Brer) Rabbit. IM thought: "I lay fretting over my identity. I suspected that I was really playing a game with myself and that they were taking part. A kind of combat" (Ellison, 1947, p. 242). He continues,

> I had no desire to destroy myself even if it destroyed the machine; I wanted freedom, not destruction…. I could no more escape than I could think of my identity. Perhaps, I thought, the two things are involved with each other. When I discover who I am, I'll be free. (Ellison, 1947, p. 243)

This is a profound statement, for when anyone discovers who he or she is, and only then, does freedom become a possibility. And since his identity exists as part of the collective identity of Afrikan people, his freedom would be complete when he united or reunited with a group, and assumed a place within it. The medical staff is concerned that he has come through the experience still feeling so strong, since the procedure was supposed to weaken the patient. Nonetheless, they decide to release him. He goes back to the personnel department and meets with a company executive, who promises that IM will be compensated for his injuries, but that he can no longer work at the plant. As he leaves the factory in a daze, IM realizes that he has changed: he is no longer afraid: "Not of important men, not of trustees and such; for knowing now that there was nothing which I could expect from them, there was no reason to be afraid" (Ellison, 1947, p. 249). IM had come to terms with the power possessed by White America, and his place within that system. The release of fear is a step toward IM assuming a new identity.

In chapter 12, a dazed IM returns to Harlem after his accident at Liberty Paints. He emerges from the subway still dazed, trying to get back to the Men's House. He stumbles, and is assisted by a stranger, Mary Rambo, who takes him to her house to rest and recover. This motherly woman nurses IM back to health. She tells him that the trouble with the modern world is that no one trusts anyone. As they talk, she encourages him to work to become a credit to his race, noting that it would be the young adults who would make changes. Mary Rambo is a mother figure, and in her care, just as in the care of his own family, IM thrives. Feeling somewhat better, IM returns to the Men's House. He sees and hears someone who reminds him of Dr. Bledsoe. Enraged by the thought of what Bledsoe did to him, IM dumps the contents of a spittoon on the man, who is a prominent local preacher. After that incident, IM left Men's House and moved into a room at Mary Rambo's place. Her name, Rambo, is reminiscent of the long-suffering Afrikan caricature, Sambo (Boskin, 1986). Without work and unsure where the future might lead, IM felt that he had lost his sense of direction. "Other than Mary I had no friends and desired none . . . . I had no contacts and believed in nothing" (Ellison, 1947, pp. 258-259). Once again he asks himself,

> Who am I, and how had I come to be? . . . Somewhere beneath the load of the emotion-freezing ice which my life had conditioned my brain to produce, a spot of black anger glowed . . . I was wild with resentment but to much under self-control,' that frozen virtue, the freezing vice. And the more resentful I became, the more my old urge to make speeches returned . . . . I became afraid of what I might do. (Ellison, 1947, pp. 259-60)

The question of who he is and how he came to be that is further evidence of how lost (out of rhythm, moving toward a negative pole) IM was.

Frustrated with reading books and pondering his fate, IM takes to the streets in chapter 13. In the cold of the day, he encounters a man selling yams. Yams are another memory from IM's childhood, his communal, Afrikan past. As he bites into one, a surge of homesickness almost overcomes him. Initially ashamed of his poor past as reflected in his taste for "soul food" such as yams (Major, 1970), he quickly makes peace with his past, saying that yams were- ". . . my birthmark . . . I yam what I am!'" (Ellison, 1947, p. 266). He wonders: "What and how much had I lost by trying to do only what was expected of me instead of what I myself wished to do?" (Ellison, 1947, p. 266). IM walked down the street, and encountered the eviction of a very old couple of Afrikan descent from their apartment in progress. He is moved to speak on behalf of the couple to the people who had assembled in front of the apartment building. The couple's plight stirred in him the image of his mother working hard when he was a child. His speech motivated the crowd to attack the men forcing the eviction and to move the couple back into their home. This resulted in violence that he both welcomed and feared. IM saw a group of White people watching the proceedings, who said that they are there because they believed in "brotherhood" (Ellison, 1947, p. 282). Soon the police arrived. With the advice of a White woman, IM escapes by going up to the roof, crossing all the roofs on the block, and descending to the street through the last of the buildings. As he moves away from the scene, one of the White men catches up to him, and praises IM for the persuasiveness of his speech. They go to a restaurant for coffee and discussion. The man analyzed him, saying: "You have not completely shed that self, that old agrarian self, but it's dead and you will throw it off completely and emerge something new. *History* has been born in your brain" (Ellison, 1947, p. 291, *italics in original*). Here a White man tells IM what his identity is. This identity is one where IM leaves his past and is headed to a new, uncharted future. IM's admirer uses a historical analysis of race similar to that of the American Communist Party (Padmore, 1972) to point out to IM that people of Afrikan descent in Harlem need a leader and that the admirer who was talking with him, named Brother Jack, was part of an organization. He invited IM to join that organization.

IM decides to join the organization in chapter 14. He meets Brother Jack and rides with him to a meeting in a wealthy section of New York at an expensive building called the Chthonian. "Chthonian" is a word that is defined as concerning underground spirits and gods (Merriam-Webster's Collegiate Dictionary, 2001). At the meeting IM is introduced to members of the organization, which is called "the Brotherhood." As IM begins his training in what he is to do for the Brotherhood, Brother Jack asks: "How would you like to be the next Booker T. Washington?'" (Ellison, 1947, p. 305), which flatters IM. Brother Jack suggests that Washington had come out of nowhere to represent his people, just as IM had done that day at the eviction. IM was to move into an apartment that the Brotherhood provided for him. Brother Jack said that he would be given a new name, and IM was told not to communicate with his family back home. IM is moving further away from his family, his community, and his culture. His identity is still swinging toward the negative pole. The reader never learns what IM's new name is, just as his original name is a mystery. He dances with a White woman for the first time, thinking to himself: " . . . white folks seemed always to expect you to know those things which they'd done everything they could think of to prevent you from knowing" (Ellison, 1947, p. 315). He was given money to buy new clothes and to pay the back rent that he owed to Mary Rambo. He thought about Mary and what he disliked about people like her: ". . . they usually think in terms of we' while I have always tended to think in terms of me'" (Ellison, 1947, p. 316). The concept of "we" is a classic statement of community and of family. IM's dislike for these things is a further indication of how culturally lost he is.

In chapter 15 IM spends his last night at Mary's place. When he awakes, he finds a cast-iron bank in the form of a Sambo character. He accidentally breaks the bank, and puts the pieces and the money that was in the bank in a bag, and puts the bank inside his briefcase. Breaking the bank could have meant that IM was breaking free of his subservient mentality, but that proved not to be the case. He says goodbye to Mary, and starts his first day with the Brotherhood. He tries twice to throw the bank and coins away, but each time a circumstance results in him retaking possession of the bank. Symbolically, the inability to get rid of the bank indicates that he cannot rid himself of his subservience.

He begins his work with the Brotherhood in chapter 16 by speaking at a community meeting in Harlem sponsored by the Brotherhood. As he walked out onto the stage, IM realized that he was to become a different person, a different personality when he began to speak. In his speech, IM tells the gathering that people of Afrikan descent in Harlem have been oppressed in many ways, and that why much of that happens is because: "*We let them do it!*" (Ellison, 1947, p. 343, [*italics in original*]). He goes on to say that he felt more human, and that he had come home: "With your eyes upon me I feel that I've found my true family! My true people! My true country!" (Ellison, 1947, p. 346). The audience responded enthusiastically to his words, but, unfortunately, what he said was not true. He was not part of Harlem; he was part of the Brotherhood. He lived outside of Harlem, and had no close connection with anyone in that community. After the speech, leaders of the Brotherhood critiqued it. Most criticized it as not being scientifically sound. It was decided that IM undergo a period of intense study of the Brotherhood's scientific theories and practices on politics so that he would be able to guide the people of Afrikan descent's excitement into the proper forms of action. IM believed that as a Brotherhood speaker, "I would represent not only my own group but one that was much larger" (Ellison, 1947, p. 353), meaning much larger than persons of Afrikan descent. This belief was another mistake. Not being centered in the culture, thoughts, and emotions of his people meant that IM was lost, even though he believed that he was on solid ground. IM also realized that the problems he had encountered with Norton and Bledsoe led him to this important place: it was his destiny.

In chapter 17, IM was ready to begin his work for the Brotherhood after four months studying ideology and ideological attitudes. Brother Jack encouraged IM to: "Say what the people want to hear, but say it in such a way that they'll do what we wish ". . . . Act first, theorize later . . . .'" (Ellison, 1947, p. 359). IM is appointed as chief spokesperson for the Brotherhood in the Harlem area. His job is to keep the community active and to increase membership in the organization. IM is introduced to Tod Clifton, a young man of Afrikan descent who is in charge of the youth in the Harlem area. IM is also instructed about Ras, a leader in Harlem who is described as a wild man and a black nationalist by Brother Jack. Brother Jack warns against any violence in encounters with Ras, saying that the Brotherhood was against violence. Clifton speaks admiringly of Marcus Garvey's ability to move his people to action. IM decides to take the Brotherhood's message to the people by speaking on a street corner as he had seen Ras do (and as Garvey had done and Malcolm X would do). When IM does so, Ras (also called Ras the Exhorter) and his men attack the Brotherhood members. They fight, and Ras gets the best of Clifton. As he prepares to kill Clifton, Ras stops, saying that he could not kill Clifton because he is of Afrikan descent, like Ras. As he exhorts Clifton to leave the Brotherhood, he claims that real brothers are the same color, and that the White persons in the Brotherhood will sell them out. IM hits Ras, getting the best of him. Ras asks: "What kind of black mahn is that who betray his own mama?"" (Ellison, 1947, p. 371). Ras goes on to say that the White leaders of the Brotherhood betray Clifton and IM, and Clifton and IM betray their people. Clifton remarks that in order to live the life that Ras suggests, one must ". . . plunge outside history . . . . Otherwise he might kill somebody, go nuts'" (Ellison, 1947, p. 377). In fact, to have done as Ras suggested would have placed IM and Clifton *back into* history, since they would be part of the history of peoples of Afrikan descent struggling for their freedom and humanity. The next morning, IM met the Harlem office's custodian, Brother Tarp, an older man of Afrikan descent, who had put a picture of Frederick Douglass[16] up on the office wall. The work in the Harlem community went well, and IM came to recognize within him two selves, the old self of Bledsoe and Brockway, and the new self as a spokesperson for the Brotherhood. He saw the power of words, which had made Frederick Douglass and himself leaders. And all the while, the reader knows that this power of words is illusory, since the most important words, his name, are missing.

In chapter 18, IM receives an anonymous letter advising him— "*Do not go too fast*" (Ellison, 1947, p. 383, [*italics in original*]). He asks Brother Tarp who had written the letter, but Brother Tarp did not know. Tarp then gives IM a link to a chain from shackles that Tarp had worn for nineteen years as member of a chain gang in the South. Tarp does this believing that the link contained a lot of significance, as breaking it had led to Tarp's freedom. It reminded IM of a polished link he had seen on Dr. Bledsoe's desk at the college. Both chains symbolize enslavement, and forms of freedom. Tarp's is the freedom of an escapee; Bledsoe's is the freedom of voluntary servitude. Neither chain offers true freedom. IM is visited by a member of the Brotherhood committee, Brother Wrestrum, who objected to the chain link being visible on IM's desk, because everything had to represent things that Brotherhood members had in common. Wrestrum suggests that the Brotherhood needed symbols of what they stood for, such as a flag and an emblem. While Wrestrum is in the office, IM is telephoned by a writer for a magazine who wants to interview him. IM repeatedly declined, recommending that they interview Tod Clifton, who had been with the organization longer. The magazine representative continues to insist, and IM finally agrees to do the interview. Two weeks later, Wrestrum brings IM to the Brotherhood's governing committee on a charge of ". . . using the Brotherhood movement to advance his own selfish interests" (Ellison, 1947, p. 400), because he did the magazine interview. IM responded that the work was going well, which proved that he had placed the work and the community ahead of himself. The committee finds IM innocent any transgression related to the magazine article, but guilty of self-promotion. IM is given a choice: leave the organization or be reassigned to lecture in downtown New York on the subject of the equality of women. He decided to accede to the reassignment. Now it is the Brotherhood that is keeping him running.

In chapter 19, IM is enticed by one of the women of European descent at a rally to further discuss issues he had raised in his speech. IM ends up sleeping with her, and is still in her bed when her husband returns. IM dreaded being called before the committee for this transgression of organizational discipline, but none occurs. A short while later, IM is called to a committee meeting, where he is informed that Tod Clifton had disappeared from the Harlem area. IM is told to return to Harlem, which is in crisis. Clifton, who had replaced IM as the Brotherhood's leader in Harlem, had not held the advantage that IM had gained for the Brotherhood with the Harlem community. IM is running again.

IM finds Harlem to be a different place when he returns there in chapter 20. He is greeted with hostility when he enters a bar in search of a member of the Brotherhood. IM learns that many of the persons in the community who had initially gotten jobs through the efforts of the Brotherhood had lost those jobs because the Brotherhood had shifted its focus away from the interests of the Harlem community. When IM got to the Harlem district office, Brother Tarp was gone as well. IM called headquarters to get information, but no one responded. He realized that a committee meeting was in progress and that they had not notified him about it. Angered, he went out and bought a new pair of shoes in another part of the city (perhaps so that he could continue running). As he walked to the subway, he was attracted by a nearby crowd. When he got to the front of the crowd, he saw Tod Clifton on his knees, manipulating a Sambo doll made of paper. Clifton was selling the dolls to people of European descent in his audience. Enraged by Clifton's betrayal of people of Afrikan descent, IM spit on the doll and then left the area without talking with Clifton. IM soon sees a police officer pushing Clifton, who responds by hitting the officer. In response, the officer shoots Clifton, killing him. As IM walked away from the scene of Clifton's death, he wondered: "Why should a man deliberately plunge outside of history and peddle an obscenity . . . "

(Ellison, 1947, p. 438), a question that remains unanswered. When IM returned to Harlem, no one acknowledged him: he, too, had plunged outside history. And like Clifton manipulating the Sambo doll, IM slowly begins to wonder if the Brotherhood might be manipulating him against his people in Harlem as well. Was he, and were the people of Harlem, Sambo dolls who were controlled by the Brotherhood?

In chapter 21, IM realized, when he got back to the district office in Harlem, that he should have kept his head and denounced the dolls and Clifton to educate the crowd. IM had taken one of the dolls, and realizes that it was controlled by a string, which had been invisible. Here is another symbol of the invisible control that people of European descent (those to whom Clifton sold the dolls) appeared to have over people of Afrikan descent. IM organizes a public funeral for Clifton at which he speaks. During the speech, IM says: "His name was Tod Clifton and he was full of illusions. He thought he was a man when he was only Tod Clifton" (Ellison, 1947, p. 457). After the funeral, IM returned to the Harlem district office, beginning to organize the emotion that community felt over Clifton's death.

In chapter 22, IM is called to a meeting of the Brotherhood committee. He is criticized for taking personal responsibility for organizing the funeral ceremony rather than acting on organizational priorities. Among his critics is Brother Tobitt. Brother Jack said that due to IM's actions, ". . . a traitorous merchant of vile instruments of anti-Negro, anti-minority racist bigotry has received the funeral of a hero" (Ellison, 1947, p. 466). IM said that the community had said that the Brotherhood had betrayed them. IM asks Brother Jack whether he sees himself as the community's ""great white father"" (Ellison, 1947, p. 473). Brother Jack responds that he is their leader. Brother Jack gets so angry that his artificial left eye pops out. The committee orders IM to attend further training. IM is beginning to become conscious of the fact that he is being manipulated like a Sambo doll by the Brotherhood. They continue to keep him running.

In chapter 23, IM is accosted by Ras the Exhorter and his men when he goes out into the streets of Harlem. Ras blames IM and the Brotherhood for Clifton's death. IM escapes from the crowd, goes into a store and buys sunglasses and a hat to disguise himself and thereby avoid further violence by Ras and his followers. At this point, IM is perceived as an enemy by the Afrikan people of Harlem; he has betrayed his birthright. When he goes outside, he is repeatedly mistaken for a man known as Rinehart. IM learns that Rinehart is a numbers runner, a gambler, a pimp and a reverend. IM happens upon Rinehart's church, where he sees a leaflet that proclaims "Behold the Invisible" (Ellison, 1947, p. 495), a symbolic reminder of IM's own invisibility and Rinehart's ability to change personas within the same community and maintain each separate identity. In his disguise, IM is no longer recognized by Ras, who has changed his name to "Ras the Destroyer" (Ellison, 1947, p. 485). IM goes to Brother Hambro, the Brotherhood's teacher. Hambro tells IM that the members of the Brotherhood in Harlem would have to be abandoned, sacrificed. Brother Hambro goes on to say: "That it's impossible *not* to take advantage of the people'" (Ellison, 1947, p. 504, [*italics in original*]). IM responds by saying: "Everywhere I've turned somebody has wanted to sacrifice me for my good-- only *they* were the ones who benefited" (Ellison, 1947, p. 505). It dawned on IM after he left Brother Hambro that he could be invisible to his own people as well. He thought: "Well, I *was* and yet I was invisible, that was a fundamental contradiction" (Ellison, 1947, p. 507, [*italics in original*]). As he pondered the possibility that he had to become like Rinehart to assist his people, he felt that he was finally accepting his past and his people. "They were me; they defined me. I was my experiences and my experiences were me . . . ." (Ellison, 1947, p. 508). He saw Norton, Brother Jack and Emerson blend into one person in his mind- White men who looked at him as a thing to be used, much like slaves had been used, and as the Sambo doll had been used. At that moment, he recognized his invisibility. He vowed to take his grandfather's advice and overcome persons of European descent by smiling and yessing them to death. He decided to appear to agree with whatever the Brotherhood wanted him to do. At last, IM was starting to reclaim his identity. His first step was to understand the ancestral dream of his grandfather's instructions. He was going to work for Afrikan people, and against anyone who did not represent the interests of Afrikan people.

In chapter 24, IM began implementation of his plan. A series of disruptions had occurred in the Harlem community when he reached the district office. In line with his plan, he falsely reported to the Brotherhood committee that he was organizing a clean-up campaign, and handed them new memberships that were fake. In an attempt to learn the inner workings of those in control of the Brotherhood, IM meets a woman, Sybil ("sibyl" meaning prophet); (Merriam-Webster's Collegiate Dictionary, 2001), at Brother Jack's birthday party at the Chthonian Hotel. He takes her to his apartment and they both get drunk. She wants him to dominate her sexually, which he sees as her worshiping a kind of power. He does not consummate the act; rather they both fall asleep. He is awakened by a phone call from the Harlem district office, saying that there was trouble there. IM sends Sybil home in a cab, and continues on to Harlem on a bus. When he gets to 125th Street, the symbolic center of Harlem, he begins to walk east from the bus toward the center of Harlem. As he walks under a bridge, birds defecate on him, a foreboding of what he is about to encounter.

In chapter 25, IM begins to hear gunfire as he reaches the outskirts of Harlem. He encounters bizarre scenes, such as men pushing a safe down the street, pursued by the police. The officers shoot at the men, and IM is grazed in the head by a bullet. He falls in with a group of men, who steal some clothes and alcohol. The group debates what started the uprising, some saying it was a response to the shooting of Tod Clifton, while others attributed it to Ras the Destroyer. With one of the men, Dupre, in the lead, they break into a hardware store and arm themselves with buckets filled with coal oil and matches. As they return to the street, they are passed by a group of men pulling a Borden's milk wagon[17], on top of which sat a woman drinking and dispensing beer. IM soon learns that Dupre and his friends plan to burn down the building in which they live because it is in horrible condition and is uninhabitable. IM realizes that these men, part of the masses of people of Afrikan descent that he had been leading, could do things for themselves: "They organized it and carried it through alone; the decision their own and their own action" (Ellison, 1947, p. 548). IM assists the men in pouring the coal oil throughout the building. As IM comes out of the building, a woman recognizes him as the Brotherhood leader and calls him by his Brotherhood name. (Again, we do not "hear" the name.) One of Ras's men hears her, and pursues IM. IM gets lost in the crowd and escapes: "I was one with the mass" (Ellison, 1947, p. 550). At this point, the Tehutian pendulum is beginning to swing back in a positive direction for IM. He is one with members of his racial group once again. But it is short-lived.

As IM moves on, helmeted and armed police appear, and

bricks are rained down on them from the tops of the buildings. One of the men in the group says that if this is a race riot, he wanted to be where people of Afrikan descent were fighting back. These words bring IM further clarity. He wondered if the Brotherhood committee had planned for this to occur. He came to view the actions of the Brotherhood as intentionally stirring up this community, committing murder by manipulating the people of Harlem into fighting against the guns of the police with bricks. Further, he saw that he had been a tool in service of the Brotherhood's plan. He continued to run, again evoking the letter in the dream: "Keep This Nigger-Boy Running" (Ellison, 1947, p. 33). He comes upon seven white female mannequins hung above the street. These mannequins are a reversal of the lynchings of men of Afrikan descent in the South, often for looking at women of European descent in the wrong way during and after the period of enslavement (Wells-Barnett, 1969; White, 1969). Ras and his men are coming down the street with guns and rifles, with Ras on a horse, dressed in Afrikan garb, evoking images of Marcus Garvey in his full regalia (Hill & Blair, 1987). IM is unable to put on his disguise; his sunglasses have been broken. Therefore, he is now not totally invisible to his own people.

Caught between the police and Ras, IM moves toward Ras, armed only with Brother Tarp's chain link, which he holds over his knuckles. Ras's men recognize IM and identify him with the Brotherhood. He responds: "I am no longer their brother . . . . They want a race riot and I am against it. The more of us who are killed, the better they like--'" (Ellison, 1947, p. 557). He is interrupted by Ras, who calls for IM to be lynched as a traitor to Afrikan people. IM continued to speak, pointing out how Ras and he were pitted against each other by the Brotherhood and other people of European descent in power. But Ras and his men are incensed and are not listening to him. IM concluded that those White men and Ras were equally blind, and that he was invisible to all of them. Ras attempts to injure IM with his spear. IM throws Ras's spear back at Ras, hitting him in the face and locking his jaws. IM escapes, running toward Mary Rambo's place. He hides himself behind some bushes to rest, and while he does so, he comes to believe that Brother Jack, Tobitt, Wrestrum and the others would soon arrive to capitalize on the confusion. He resolves to attack them, and starts toward where he expects them to be. He is confronted by some men of European descent who want his briefcase, which has been his constant companion since he received it at the battle royal. As he runs from the men, he falls into an open manhole and lands on a pile of coal. The men replace the cover on the manhole, leaving IM in ". . . a kind of death without hanging, I thought, a death alive" (Ellison, 1947, pp. 566-567).

Unable to see, and unable remove the manhole cover, IM is forced to burn some of the papers in his briefcase in order to see. He burns his high school diploma, then Clifton's doll, then the anonymous letter that had told him not to move too fast in Harlem, and then the envelope with his Brotherhood name on it, destroying each of his old identities at the same time. He realizes that the handwriting was the same on both the letter and the envelope, and that the handwriting was that of Brother Jack. "That he, or anyone at that late date, could have named me and set me running with one and the same stroke of the pen was too much" (Ellison, 1947, p. 568). IM was furious and outraged. Eventually he made peace with all of it, saying to himself: "That's enough, don't kill yourself. You've run enough, you're through with them at last'" (Ellison, 1947, p. 568). He falls asleep and has a nightmare, another form of dream. In it, he is imprisoned by Jack,

Emerson, Bledsoe, Norton, and his school superintendent, and others, all of whom had kept him running. He rejects them and they respond by castrating him. IM laughs at them, now freed of illusion. He tells them that not only was his future dead due to the castration, but so were they. He says that the blood they hear dripping "*. . . is all the history you've made, all you're going to make*" (Ellison, 1947, p. 570, *italics in original*). Then IM awakes: "And I awoke in the blackness" (Ellison, 1947, p. 570), which is both a literal and symbolic statement: the space he was in was dark, but now he was once again his Afrikan self. He now felt whole. He believed that he could no longer go back to Mary, to the campus, to the Brotherhood, or home to his family. His only choices, in his mind, were to move ahead or stay in the underground. He decided to stay in the underground. He concludes: "The end was in the beginning" (Ellison, 1947, p. 571), which leads to the prologue.

Invisible Man: The prologue and the epilogue. The prologue and the epilogue take place underground, evoking short novels by Dostoyevsky (1864/1948) and by Richard Wright (Wright, 1944/1969), which take place underground as well. IM begins by saying: "I am an invisible man" (Ellison, 1947, p. 3). He goes on to explain: "I am invisible, understand, simply because people refuse to see me" (Ellison, 1947, p. 3). He says that he is invisible because those he encounters fail to see him. An example of this was a man who bumped into IM on the street. The man cursed IM and in response, IM beat the man badly. He had gotten to the point that he fought back against those of European descent who did not see him or acknowledge his existence. One way he fights back is to rob them of power: electrical power, which he is draining off from the power company to use to light his underground abode. He also used the electrical power to play music, especially the blues. He says that in his twenty years of living, he had not become alive until he realized that he was invisible.

One night he smokes some marijuana, and under its influence, he dreams that he hears an old woman of Afrikan descent singing a spiritual, which evolves into visions of levels of oppression suffered by persons of Afrikan descent. He finally arrives at a level of this dream where a sermon is being delivered by a preacher of Afrikan descent that is entitled: "the Blackness of Blackness'" (Ellison, 1947, p. 9). The vision evolves and IM finds himself in conversation with the old woman who was singing the spiritual. She explains that her master had impregnated her four times, resulting in the birth of four sons. The master had promised to free them all, but had failed to do so. IM asks her what the freedom was that she desired, but she could not answer. IM next encounters music, and hears footsteps coming up behind him. IM calls out, believing that either Ras the Destroyer or Rinehart is chasing him. His vision recedes, and he hears Louis Armstrong singing "What did I do To be so black And blue?" (Ellison, 1947, p. 12). He then comes out of the dream. The spiritual, the sermon, and the woman's story all are pieces of Afrikan history and culture. In the dream, he is both a participant in and an observer of this history, a history of which he is already aware. The dream symbolizes his rebirth. He concludes that, despite the tribulations he underwent in the Brotherhood, action is necessary. He says: "A hibernation is a covert preparation for more overt action" (Ellison, 1947, p. 13). He ends by saying that he should have really hurt the man who bumped into him on the street. He is growing more assertive about who he is. IM then tells the stories of his life that occupy the next 25 chapters.

In the epilogue, IM reviews what he has learned. He realizes that: "I was never more hated than when I tried to be honest. Or when, even as just now I've tried to articulate exactly what I felt to be the truth" (Ellison, 1947, p. 573). He also says that in trying to do what others wanted him to do rather than what he wanted to do himself, problems resulted. Then he again tries to make sense of his grandfather's dying declaration, but is again confused. IM tries to figure out what he desires, but cannot figure it out. He knows it is: "Certainly not the freedom of a Rinehart or the power of a Jack, nor simply the freedom not to run" (Ellison, 1947, p. 575). But, since he could not figure out what he did want, he had stayed underground. He knows that a part of his difficulties is in the world around him, and a part is in him. He believes that his life is one of infinite possibilities, though he knows that society presses conformity onto its citizens, a status he rejects. "Why, if they follow this conformity business they'll end up by forcing me, an invisible man, to become White, which is not a color but the lack of one" (Ellison, 1947, p. 577). His belief is that each race should be allowed to be what it is. People of European descent should not try to become more Afrikan, and people of Afrikan descent should not try to become more European.

He then recalls seeing Mr. Norton lost in the subway one day. IM reminded Norton that Norton had said that he, IM, and the other students at the college would create Norton's destiny by living their lives. IM asks if Norton is ashamed of his destiny, thinking to himself about who he (IM) is, and what he has become. Norton does not recognize him and believes IM is crazy. Thereafter, IM periodically gets the desire to return to the South, but believes that the answers to his desires are in his own mind. Had he returned to the South, he would have had the opportunity to resume his place within his family, his community, and his culture. IM concludes that he wrote all of these events down because in so doing, he confused himself, thereby ridding himself of some of his anger and bitterness. He realizes that his grandfather, though enslaved, never had any doubts about his humanity like, he (IM), his offspring, does. And IM decided to end his hibernation and return to the world. He believed that he had overcome all of his obstacles, ". . . except the mind, the *mind*" (Ellison, 1947, p. 580, *italics in original*), evoking the first Tehutian principle, All is Mind/spirit. IM concludes the epilogue by saying: "And it is this which frightens me: Who knows by that, on the lower frequencies, I speak for you?" (Ellison, 1947, p. 581). By ending the novel in this way, IM and his search for who he is as a male of Afrikan descent in America, becomes an invitation for the reader to search for who she/he is. He has found his identity. He knows that he is Afrikan and part of a vibrant community. He leaves the reader to speculate what he will do, and where he will go. But the reader is clear that he is no longer running from himself or from White people. He is his own man. As such, he is capable of playing a role that benefits communities of Afrikan descent, as well as himself.

Invisible Man: Criticism and Critique

Introduction. IM says that he is twenty years old when the events in the novel take place (Ellison, 1947, p. 15). He is a young man, and as such, he does not really know what he knows about himself: he is still searching. In analyzing his character and the book, much depth can be attained, but that would not serve the purposes of this discussion. Rather, the events in the novel, their consequences for IM, and his understanding of them will be used to understand the development of his identity as a male of Afrikan descent in the United States, and how such literary characters can fill in emerging Afrikan-centered male identity development theory.

In order to proceed with this discussion, it is useful to place *Invisible Man*, that is, to locate it in the historical period in which the novel takes place. While there is no date given for any of the events, one can draw inferences from events and occurrences in the novel. The prologue, we learn, takes place 85 years after his grandfather learns that he and other slaves have been freed from slavery. The Emancipation Proclamation freed slaves in the South in 1863, and the Thirteenth Amendment to the Constitution of the United States, which abolished slavery, was passed in 1865 (Franklin & Moss, 2000). IM is approximately twenty years old when he speaks from the "present" of the novel (Ellison, 1947, p. 15) about his life over the past year or so. This places the "time" of the novel in the 1940s, during or just after World War II. There are other clues that place the time of the novel in the 1940s, such as the Borden's milk wagon, the references to subways and buses in New York, the use of coal oil to fire heaters in apartment buildings in the winter and fans to cool them in the summer, and the use of submachine guns to defend a bank. There are also references to famous persons of Afrikan descent from the past: Frederick Douglass, Booker T. Washington, and Marcus Garvey, and from the present of the novel, Joe Louis, Paul Robeson, and specifically Earl Sande, one of the few jockeys of Afrikan descent who raced horses in America during the era from the 1920s to the 1950s *(Saluting all-time greatest jockey*, 2005). Ellison actually competed the writing of *Invisible Man* in the 1940s. Ellison was consciously using markers in Afrikan American history, such as the emancipation from enslavement, Jim Crow laws that forced IM and the vet to ride in the back of the bus in the South (Woodward, 1974), references to Booker T. Washington and Marcus Garvey, and to Earl Sande, to indicate that an Afrikan sense of time, one marked by events in Afrikan American history, was being utilized in the novel.

Another aspect of this novel that requires clarification is the generations, by age, that are referenced in the novel. IM repeatedly talks about and remembers his grandfather. Other members of the group that is two generations older than IM are the Founder, Mr. Norton, Dr. Bledsoe, Reverend Barbee, and Mr. Brockway. Those in the generation immediately above his are his father and his mother, who he remembers as he ponders his fate in college and when he is in Harlem. Also in this generation are Mary Rambo, Brother Jack, and the vet. In his generation are Ras and Tod Clifton. He never mentions siblings or any other family members. The reader learns little of his family. He appears to be disconnected from his kinship network, and in that sense, has separated himself from his racial and cultural group. In short, he has lost himself.

The Fields of Literary Criticism and Critique

The western literary world has long used psychological theories, analyses, and applications to explain the inner workings of literary characters (Natoli, 1984). Novels that focus on the development of the emotional lives of the characters are called "psychological novels" by those who construct and define mainstream literary terms (Beckson & Ganz, 1989; Thrall, et al., 1960). For example, a long-standing journal in the field of literary criticism, published from 1951 to 2004, was entitled *Literature and Psychology* (Modern Language Association of America). There is also a book by the same name (Lucas, 1957) that covers the same subject area. Edel (1982, p. xv) uses the term "literary psychology" to discuss the psychological motives for authors to write what they write (see also *Shakespeare's Personality* (Holland, et al., 1989, and *The Modern Psychological Novel* (Edel, 1964), which utilize the same framework). The focus of the present analysis is on a character in a novel, and not on the author. It is worth noting that Edel (1982) speculated that Ellison did not write any novels after *Invisible Man* because he had totally healed himself of his internal angst, and therefore had no more to say and no need to create further fantasies to express himself[18].

In recent decades, literary critics have begun to compare authors of European descent with authors of Afrikan descent, using the tools of psychology to perform the analysis (Eichelberger, 1999; Moreland, 1999). Recently, some authors have taken psychological theories and used them to analyze novels written by novelists of Afrikan descent (Mbalia, 1995; Tate, 1998). And some authors have focused solely on the psychological issue of identity in novels by persons of Afrikan descent (Ginsberg, 1996; Lock, 1994). This present discussion reverses this analysis. Here literature and literary characters created by and about persons of Afrikan descent are used to explain and expand psychological theory, specifically Black and Afrikan-centered psychological theories around identity development for males of Afrikan descent.

A psychological novel, one that delves into the emotional lives of its characters (Holland, et al., 1989) is to be distinguished from a historical novel, one that uses history to frame or reconstruct events (Beckson & Ganz, 1989). A political novel or propaganda novel is one that deals with political or sociological issues, and offers a solution (Thrall, et al., 1960). And while European forms of literary definition compartmentalize these forms, so that each work is categorized according to its primary form, some novels, such as *Invisible Man*, combine these types.

Another concern that informs literary analysis is aestheticism, the appreciation of beauty (Merriam-Webster's Collegiate Dictionary,

2001). This concept, which has popularly been described as "art for art's sake'" (Beckson & Ganz, 1989, p. 5), posits that art should exist independent of any political, social or moral imperatives. Out of this school of thought came the concept of "aesthetic distance." "Aesthetic distance" is a literary device used by an author to separate the reader from what is being read, so that the reader does not confuse the work of art with real life (Beckson & Ganz, 1989; Thrall, et al., 1960). This idea has gained currency in American culture today. For example, many adults are currently concerned that adolescents are confusing hip hop lyrics, movie characters, music videos, and video game scenarios with real life experiences and values, which the young then emulate and imitate, resulting in harm to themselves and others, according to the theory (Kitwana, 2003; Long, 2005; Rodriguez, 2003).

The foregoing discussion of the psychological novel, aesthetics and aesthetic distance is significant because literary critics of Afrikan descent were rarely, if ever, consulted or included in the discussion. Criticisms of *Invisible Man* which emanate from European sources, or from Afrikan critics with European sensibilities, misdirect the analysis, in much the same way that Bledsoe and Brother Jack misdirected IM. The main focus of these critics is to use European forms and standards by which to judge Ellison's creation. In so doing, they make the same transubstantive error[19] that mainstream psychologists make when theorizing on people of Afrikan descent. They see IM and the other Afrikan characters as pawns, victims of the White characters in the novel. All their analysis comes from the perspective that Whites act and control, while characters of Afrikan descent react and accept their fate.

However, with the advent of the Black Arts Movement in the

1960s, that began to change. Beginning at about the same time as the advent of the Black Power movement and the advent of Black psychology, but preceding them, the Black Arts Movement added a political and social focus to the "art for art's sake" aesthetics of the era (Jones & Neal, 1968; Major, 1968). "The Black Aesthetic," as this focus was called, was a framework by which art created by persons of Afrikan descent, could be evaluated, both in terms of its artistic relevance, but also in terms of its social relevance (Karenga, 2002). The seminal work in the field, Gayle's *The Black Aesthetic* (1972), brought together theorists who defined this movement, as well as those who discussed it in terms of music, poetry, drama, and fiction. It is this distinction between a purely aesthetic critique of art, and a socially informed *and* aesthetic critique, that has been the focus of the analysis of Ellison's *Invisible Man* by critics of Afrikan descent since its publication. Both Ellison and his novel have been analyzed from the Black Aesthetics perspective. History has a way of softening the rough edges of reality, and such has been the case in the analysis of Ellison and *Invisible Man* in recent years (Butler, 2000; Callahan, 2004; Jackson, 2002; Warren, 2003).

This discourse uses both the "art for art's sake" and the Black Aesthetics frameworks for its analysis of Ellison's *Invisible Man*. This is done as a means of understanding how IM's journey assists in identifying issues of importance in Afrikan male identity formation. The Black Aesthetics framework will be modified to include the criteria enunciated by Karenga in his book, *Kawaida Theory* (1980). There he states that all creativity must be functional, collective, and committing to achieve the social relevance that Black Aesthetics demands. According to Karenga, a functional art has social relevance to persons of Afrikan descent, as well as a purely aesthetic purpose and goal. A collective art comes out of the experience of persons of Afrikan descent, and is expressed in idioms that persons of Afrikan descent use and can understand. It invites the reader to feel the creation, and to expand their understanding of and pride in themselves and their community. A committing art is affirming art, pointing persons of Afrikan descent toward the future, freedom, and liberation from forms of oppression. Writing from a continental Afrikan perspective, Chinweizu, Jemie, and Madubuike (1983) echo Karenga's view, saying that Afrikan literary criticism is, at the same time, a social criticism. They contend that literature from the Afrikan continent, as well as its criticism, must move to reject European forms and opinions. They contend that it is the social value of the work, the impact on the consciousness of the society or community in which it operates, that is important. Asante (1980) concludes that one cannot be a writer who just happens to be of Afrikan descent, because to do so denies people of Afrikan descent their heritage and their identity. Kent (1972), on the other hand, found Karenga's framework stultifying for the artist, though Kent was equally opposed to the "art for art's sake" school that removed artists of Afrikan descent from their racial and cultural community.

The Black Aesthetic and Invisible Man

One of the ways that persons of Afrikan descent communicate the symbolic content of their culture is through what is known as the "creative arts" (Gayle, 1972; Welsing, 1991). In traditional Afrikan societies, story-telling and history, which were transmitted orally from one generation to the next, have evolved, influenced by the European paradigm of knowing, into written forms of expression, such as poetry, folk tales, short stories, and novels. (Bascom, 1992; Lester, 1969; Radin, 1952/1983). The world of the novel Invisible Man is an Afrikan world. The food (chitterlings, yams), the music (blues, spirituals, Louis

Armstrong), the use of language (Black English Vernacular or Ebonics (Anderson, 1994; Dillard, 1972), the historical reference points (epoch of enslavement, sharecropping, historically Black colleges, Harlem as a "Black" city, the Harlem "riots" of 1935 and 1943 (Bennett, 1988), the involvement of persons of Afrikan descent in the American Communist Party (the "Brotherhood" in the novel) the involvement of Afrikan peoples in the Back to Afrika movement (symbolized by Ras), and the heroes of the community (Frederick Douglass, Booker T. Washington, and Marcus Garvey), all are aspects of Afrikan culture evoked in the novel. This Afrikan-centered analysis looks at the symbols, metaphors, and images in a creative work, such as a novel, in terms of their social relevance. That is, the question becomes what the creative work does to reflect the values that Afrikan people hold as central to their lives and culture. Furthermore, this work is judged by how it assists in the quest of people of Afrikan descent for liberation from oppression and for equality in the United States.

In this regard, Bone (1970) notes that the naming of persons of Afrikan descent by persons of European descent, disrupted Afrikan lineages, and thereby negated the kinship ties that were an essential feature of Afrikan cultures. IM is nameless in the novel but has not lost his Afrikan identity. He gets a name from the Brotherhood, but never finds his identity or his humanity in that context. E. M. Jackson (1970) sees but four alternatives for a person such as IM, who is working himself out of his subjugation: segregation, separation, integration, or revolt. Being nameless and invisible, IM is engaged in both symbolic separation and revolt (living underground and rejecting European roles, such as factory worker or leader). He is also thus reborn. At the conclusion of the novel, he has realized that his identity, though still forming, is made up wholly of Afrikan parts. Gone are the symbols of the Europeans. He has burned his high school diploma, his Brotherhood name, the letter warning him to slow down (a form of twentieth century shackles) and the Sambo doll, symbol of his being externally controlled. He has adopted the blues as his rhythm, the Blackness of Blackness as his spiritual framework, and the old woman's reality (that she had to kill her White master to realize her own freedom) as his credo. In short, he is no longer subjugated, because he glimpses a new Afrikan-centered consciousness that is not bounded and defined by European forms.

This provides a clear contrast to Gray (1978), who compares *Invisible Man* to Herman Melville's works *Beneto Cereno* and *Moby Dick,* and sees race as of secondary importance in the novel. Instead, for Gray, the focus is on color as symbolism. From this perspective, Whiteness is symbolic of harm throughout the book. Gray sees *Invisible Man* as an expression of the triumph of American ideas, and not the orientation of individuals. Gray sees IM's grandfather wearing a mask of humility (evoking Dunbar's poem "We Wear the Mask"); (Dunbar, 1913) to resist White control, and Bledsoe's wearing a mask of subservience for personal gain and power. In reality, neither man was allowed to be who they really were, since the white supremacist system in which they operated, prevented it. IM's grandfather was, however, operating in the best interests of his family, doing what he had to do to survive, echoing the survival thrust concept enunciated by Baldwin/ Kambon. Gray views people of Afrikan descent as invisible in the novel because of the shadow of power that Whites cast over them. This perspective exonerates persons of Afrikan descent for their plight; they are only victims and must accept their fate. In reality, people of Afrikan descent are not invisible in the novel: they are central. They either work together to advance their group interests, or work singly, mistakenly moved by an American dream that was not meant for them. Horowitz (2000) also focuses on American symbols, but does not address the symbols and images of the rich Afrikan culture that exists throughout the novel. He sees the novel as a series of White images in which the characters of Afrikan descent operate, rather than a novel placed in an Afrikan landscape, with a White background. By misplacing the text, he misjudges the symbols. Hence, he would see the optic white paint as a symbol of oppression externally imposed on the Afrikan characters, rather than seeing the irony of the black base for the optic paint, representing the work and genius of Afrikan people as the key to the success of the United States.

Rosenblatt (1974) points out a seeming irony of IM's situation: each time he accommodates-- to his college probation, to his supervisors at Liberty Paints, and to the Brotherhood-- he loses. But when he escapes the Brotherhood and Ras, ending up with nothing and living nowhere, he is most free. Warren (1974) calls that freedom a human triumph over oppression. In both cases, these critics miss the consistencies in these situations. IM is not accommodating to these situations and triumphing over oppression. Rather, he is a person of Afrikan descent in an oppressive situation seeking identity, manhood, and freedom. He makes a series of decisions, each of which has consequences. In other words, IM is an active agent in defining his own destiny, rather than a pawn or doll being manipulated by others. Some of his decisions do not work out especially well, but he survives and learns from each of them. In that sense, his is an Afrikan triumph.

The term "invisibility" has gained much currency since the publication of *Invisible Man*. Its use, among writers of Afrikan descent, connotes a number of things. Halsell (1969) quotes a nameless man of

Afrikan descent with whom she spoke, who said: ". . . when you're *born* black, you get prepared for the shame'" (Halsell, 1969, p. 19, [*italics in original*]). In other words, people of Afrikan descent are treated as though they are invisible, insulted and dehumanized by people of European descent on such a consistent and regular basis that shame becomes a way of life. These insults occur as though the person of Afrikan descent were not even there, as in the situation in the prologue where the man bumps into IM on the street as though he did not see IM there. Coley (2001) uses the term "(In)visible men" to refer to the scarcity of studies on the part played by fathers in the development of children and families. Connor and White (2006) make the same point. Franklin (2002) described the challenge that men of Afrikan descent have being recognized and acknowledged as the "invisibility syndrome." Myers, Abdullah and Leary (2000) call this a multigenerational trauma caused by individual and collective oppression. Ani (1994) called it "Maafa," the horror of the processes and results of the enslavement of Afrikan peoples over three centuries (DuBois, 1896/1969). However, that invisibility does not exist inside all persons of Afrikan descent. Rather, it is predominantly an attitude in those who do not see these persons of Afrikan descent as human beings. Mazrui (1986), speaking about the same issue on the continent of Afrika, summarized this issue by saying that power is in the hands of those who control the means of destruction.

By contrast, Schraufnagel (1973) takes the perspective that IM is both invisible to others and to himself, due to his lack of awareness of the racist world in which he lived. Schraufnagel rightly notes that the section of the novel at the college shows how people of European descent appear to be supporting education for persons of Afrikan descent, but instead are contributing to the continuation of White superiority. Schraufnagel offers interpretations for the symbols that IM carries with him when he falls into the hole near the end of the novel. These symbols are all creations of White characters in the novel, exemplifying the way they view people of Afrikan descent: as objects to be named, controlled, and manipulated against each other. These interpretations are useful, but come from the perspective that the people of Afrikan descent in the novel are victims rather than agents of their own destiny. As an agent of his own destiny, IM burns those symbols and emerges with his Afrikan identity clarified for himself as the novel concludes, and ready to re-enter the world.

Nadel (1988) and Moreland (1999) do a comparative analysis between the novels of Ellison and other writers. Moreland sees *Invisible Man* as primarily a historical novel, which is one of its many facets, perspectives. These historical events are part of the process by which people of Afrikan descent in America have asserted their desires and group goals in the American context, constantly searching for freedom. Tanner (1974) sees the conversation between IM and the vet on the bus as a key moment in the novel. The vet tells IM that his freedom will have to be symbolic as opposed to real, and that he must create his own identity. He cannot rely on the identity created for him by others. In other words, IM must be active in creating his own destiny. He must operate inside Afrikan history in America in order to create his own identity. And he must operate within the Afrikan community, from which he, his values, and his culture, spring. The vet acts as a family elder, guiding a young male toward manhood.

Gayle (1970a) enunciates a view of Ellison and *Invisible Man* held by progressive people of Afrikan descent in the Black Power era of the 1960s and 1970s. He praises the craft, education, and cultural understanding of Ellison, and suggests that Ellison was uniquely qualified to bring peace to an Afrikan American community battered by the unrelenting power of White America. In choosing to view the protagonist in *Invisible Man* as attempting to obtain validation from America proper, Gayle accuses Ellison of seeking peace at the expense of race and culture. This stance concerning emergent Afrikan selfhood in the United States is important because it insists on self definitions for Afrikan people in the face of a societal ethos that has consistently denied Afrikan existence. It contributes to the uncompromising position that wrestles a positive self-image from the stereotypes that oppress Afrikan people and make of them products of an alien worldview. However, when Gayle (1970b) equates IM and Rinehart as being without a distinct identity, he misses the point. IM is training to do, in a positive way, what his grandfather suggested: present one personality to those of European descent while he continued to develop his Afrikanity and moved toward empowerment for those of Afrikan descent.

Blake (1986) correctly notes that the central theme in *Invisible Man* is the quest for cultural identity. However, she errs in concluding that Ellison's treatment transforms this quest from one of cultural identity to an explication of the human condition. Blake bases her analysis on Ellison's juxtaposition of Afrikan American ritual alongside corresponding elements in western mythology, hence removing its distinctiveness and making it part of common human experience. She asserts that, in so doing, Ellison distorts and denies the uniqueness of Afrikan American folk expression. In reality, *Invisible Man* creates, or recreates, rather than distorts, Afrikan American folk expression. Unlike most rituals and folklore, Ellison's tale exists on the political, historic and psychological levels, *as well as* on the cultural and symbolic, the historic regions of ritual and myth (Welsing, 1991). By creating, and re-creating Afrikan American life on all those levels simultaneously, Ellison has expanded, rather than reduced, the rituals and folklore found in *Invisible Man*. Blake correctly places *Invisible Man* in the category of epic tales, describing it as a quest "for black self-definition." However, she loses her point when she states that the folk tradition that Ellison creates is a White folk tradition. In actuality, as well as in the book, White reality and Afrikan American reality exist parallel to each other, rather than one relying on the other. For example, Nichols (1970) sees *Invisible Man* as a failed pastoral dream, with the college being the pastoral and the underground the sunless, desolate end. Nichols sees a theme of "psychic emasculation" of males of Afrikan descent (1970, p. 73) operating in the novel. Nichols concludes that IM lacks social vision, and can only see himself as a pawn in a social revolution where violence appears to be inevitable and the only option. From this perspective, the Afrikan characters have no voice, and no decision-making capacity. Rather, they are manipulated by the White characters, much like the Sambo doll is manipulated. Schafer (1970) suffers from the same misgivings. He views the novel as a series of confrontations by IM with a European American society that was in charge. Nichols' and Schafer's perspectives are not useful because they make IM an object, rather than the subject of his own destiny, acted upon, rather than acting. IM acknowledges his culture throughout the novel, and embraces it in the dream sequence in the epilogue.

Morrison (1992) argues that all of the core facets of mainstream American literature, such as the focus on individualism and masculinity, are subconscious responses to the presence of an Afrikan ethos in the United States. Hence, for Morrison, the two worlds (Afrikan and mainstream American) are, in fact, intertwined. From this vantage point, oppression exists, but the response to it by members of the

Afrikan community is an action of strength. On the other hand, the response of those in control is an action of weakness disguised as power and control. Thus, it is not true, as Blake suggests, that the protagonist in the novel initially rejects the Sambo mentality, and eventually learns to accept it. At the novel's end, IM is a self-determined man.

Starke (1971) believes that IM undergoes four forms of initiation in the novel: to look at the naked White woman that men of Afrikan descent cannot touch; the battle royal fight with other males of Afrikan descent for the promise of a reward; the double-cross when the money that was the reward for fighting is not real and the carpet is electrified; and IM's speech where urging people of Afrikan descent to stay in their place wins him a college scholarship. In Starke's analysis, IM has no personhood; he is a doll in the hands of those who control his thoughts and actions. But Blake and Starke misunderstand who IM is. As he learns from experience, IM gets greater insight into who he is, both within the Afrikan world and outside it, in the context of the worlds he occupies. The self that tells the reader in the epilogue that he is now clear enough about who he is to write a book to explain himself has not accepted his Sambo-ness; he has rejected it. His desire to write his truth signifies his emergence from his underground home back into the light of reality, free, proud, and clear about who he is and who he is not. He has not acquiesced to the White world, but rather has reframed his place in the Afrikan American world, whether the rest of the world recognizes him or continues to treat him as invisible or not. The psychological frames in which this growth and transformation occur offer valuable lessons for Afrikan American people, and, in particular, those who aspire to and have achieved racial consciousness. For it provides a platform, a plane, a history, and an ideological basis from which to view and make choices in the world.

When Blake states that Ellison counterposes Afrikan American folk tradition to the doctrines of the white world, she has both overstated and understated the case. Afrikan American folk tradition is descendant from Afrikan antecedents, and therefore does not owe its existence to a comparison with White doctrine. This stance is in line with the stream of reasoning that Asante (1980) and Ani (1994) have offered. The Afrikan world, born of an Afrikan unconscious (Bynum, 1999) that continues to operate within and between persons of Afrikan descent, has a complete cosmology from which it operates (Nichols, 1970). More importantly, Ellison paints the pictures of these two seeming worlds in parallel, rather than in contrast. As Blake (1986) herself so cogently says: "The meaning is not in the thing itself but in the way it is used" (p. 89). Thus, when Blake concludes that Ellison posits that change for Afrikan Americans must occur in their minds as victims of oppression, she is correct as far as she goes. However, that is not where the issue, or Ellison's treatment of it, ends. For in changing self-definition and self-awareness, Ellison's protagonist becomes visible to himself. From this new vantage point, white oppression is no longer a force to be tolerated and survived; it is now part of the landscape of a newly beheld world, where anything is possible, and where life is lived on Afrikan American terms from Afrikan American perspectives.

Slatoff (1989) does an excellent job of outlining the psychological aspects of *Invisible Man*. He sees IM as trying to stay human by adjusting to his own anger and bitterness, his own love and hate, " . . . to cope fully with the mental and emotional damage caused by having grown up black in a white society-- the effort of such a person, that is, to stay sane" (Slatoff, 1989, p. 32). He points out the identity problems that the protagonist has because he has no name. Slatoff also notes that an individualized self presupposes that someone, anyone, cares about what that person wants or needs. However, Slatoff's conclusion is faulty. He believes that the damage done to IM by society is irreparable because he has no self on which to rely. Hence, IM is unable to become the smiling, Sambo-like man that his grandfather proposed, or to reject the Sambo character entirely, because both choices are dehumanizing. Slatoff believes that White society offered Afrikan Americans two choices: be Sambo or be dead. IM, Slatoff contends, can choose neither.

Slatoff (1989) does offer two sources of hope: the protagonist's ability to laugh and have a sense of humor, and his desire to be human, to love and to hate. His analysis is important because it points to the shape and contour of the psychological terrain of the novel, and because it posits some beginning solutions to the issues raised. However, his analysis is scarred by his perspective. IM does indeed have a sense of self. That self is as insane as are all persons of Afrikan descent who are conscious of and have to make peace with their oppressive situations in this society and still maintain their identity as people of Afrikan descent. Choosing to be accommodating (Sambo-like) or assertive (like Ras) are not linear choices. Rather, they are styles of survival. The issue is whether IM and persons of Afrikan descent generally can sustain a positive, constructive self and culture and prosper without sacrificing their core identity and striving to be acculturated.

Wolfenstein (2003) also uses psychoanalysis to discuss *Invisible Man*. He points out that people of Afrikan descent in the novel see themselves through the eyes of people of European descent, as deformed and inferior. He states that IM lacks a connection to family or friends, male or female. This view appears to be short-sighted and incorrect.

Each of the Afrikan characters has their own view of themselves, and operates in a separate world of Afrikan people where Whites are bystanders, much as Brother Jack was a bystander when IM roused the gathering as the old folks were being evicted from their apartment. IM does not lack connection with his family. Rather, he has not communicated with them. The connection is present in his memories, and in the esteem he holds for them. His isolation is a choice that he makes on the road to discovering his personhood. Wolfenstein focuses on Clifton's statement that people of Afrikan descent must occasionally go outside history to avoid going crazy and killing someone. Another way to interpret Clifton's message, is that people of Afrikan descent have to go outside of history as defined by those of European descent, and into Afrikan history as defined by persons of Afrikan descent, in order to regain and retain their racial and cultural identity. Wolfenstein concludes that this novel is not to be taken literally, but is to be taken literarily. In so doing, he minimizes the importance of *Invisible Man*, for it is to be taken literally *and* literarily, as well as historically, metaphorically, racially, and in a number of other ways. Borrowing from Dixon's (1976) term, the novel is "multiunital," many things at the same time. Wolfenstein concludes that nationalism, integrationism and communism, all European concepts, are discredited in the novel. In so doing, he imitates other psychiatrists and psychologists who utilize Eurocentric psychological theories to conclude that people of Afrikan descent are marginal (Lane, 1998). Nationalism is not discredited in the novel. Rather one form of nationalism, a one-person, charismatic, leadership driven form of nationalism in the form of Ras the Destroyer, is rejected.

Gibson (1981) asserts that the writing of the novel has allowed IM to deal with his past psychologically. This analysis echoes the

Freudian perspective that Greene (1996) uses to show that as IM breaks free from his Oedipal struggle with Bledsoe, Norton, and Jack, his supposed surrogate fathers, he frees himself. IM's "accidents," such as the verbal error in his speech as the battle royal, are seen as aspects of his personality emerging through the definitions that a repressive environment forces on him. Gibson overlooks or misses IM's own sense of his humanness as these "accidents" occur. Gibson believes that IM's individuality precludes group action, such as the nationalism that is symbolized by Ras. In fact, IM has an identity throughout the novel. It is one that is grounded in the values that his family has provided him. His identity evolves and emerges as he acts in various situations, and reacts to situations in which he finds himself. These actions are evidence of a functioning identity rooted in the Afrikan respect for the elders of his community in New York.

Baker (1986) also uses a psychological lens to examine the novel. He says that the phallus of men of Afrikan descent in the novel is a symbol of force that men of European descent envy and want to destroy. Baker's view gives agency to the males of Afrikan descent in the novel, but in his view, ultimate control still resides in the men of European descent. This view gives IM a genetic form of power, but he does not control it. Rather, Whites confer it on him by their fear of his genetic potential. This theme is a main focus in Welsing's (1991) explanation of the operation of the system of white supremacy and racism throughout the world.

Bennett and Nichols (1974) utilize Fanon's analysis from his work *Black Skin, White Masks*, which states that oppressed people live in a region where an authentic upheaval, both internal and external, can be born. These authors see the use of violence in Afrikan American fiction leading in two directions: toward self-destruction and toward self-discovery. They analyze the scene in *Invisible Man* where Tod Clifton is found selling Sambo dolls on the street, is confronted by the White police officer for selling without a license, and where Clifton is killed by the officer. In this scene, according to the authors, we see Clifton's self-destruction (selling dolls and getting killed) and his self-discovery (standing up to power and not succumbing to their authority). The authors also point to the scene in the prologue where IM discusses hitting and knocking down a White man who bumped into him as though IM did not exist, and later remarking that he should have killed the man. In accepting the meaning that can lie in death and violence, the authors conclude that such revolutionary violence had been found in Afrikan American writers for some time, and that this dimension of that fiction had perhaps been the reason that Afrikan American novels had been excluded from literary anthologies and courses on American literature. The revolutionary aspect of this portrayal of Clifton and of IM, as well as of Ras the Exhorter/Destroyer, places this novel in the tradition of Afrikan American protest literature. As such, it shows the centrality of self-definition and self-actualization as cures and correctives to dehumanizing situations. This stance has important implications for Afrikan American persons who struggle in this society. It points to the possibility that actions and attitudes that are considered pathological by the dominant culture may, in fact, be healthy and liberating to the person making the expression. In line with this view, Davis (1986) believes that IM discovering his Afrikanity is a starting point toward this liberation.

Walling (1973) says that IM's situation is summed up in his grandfather's riddle. IM must accept enslavement or seek freedom, must accept servility or seek rebellion. In Walling's view, Tod Clifton was crushed between the Brotherhood, which divorced itself from the issue of color, and his Afrikan consciousness, which was coming out when IM first met Clifton. Clifton eventually lost his mind due to the stress of these opposing forces, according to Walling's perspective. Walling sees *Invisible Man* as a journey by IM from victim to master, from a deluded boy of Afrikan descent to a "conscious selfhood, capable of drawing the power for self-knowledge from out of a seemingly oppressive social structure" (Walling, 1973, p. 14). In a similar vein, Ford (1970) views IM's acquiescence throughout the novel not just as a guarantee of survival for persons of Afrikan descent, but also as a strategy designed to destroy those who oppress people of Afrikan descent. These authors take the perspective that IM is an active presence in the novel, rather than a passive force, reacting to control of his life by forces outside his control and from outside his community.

Clarke (1970) notes that the best characters in *Invisible Man* are the ordinary people, like the veterans, Mary Rambo, Brockway, and the people IM meets during the uprising in Harlem. This is an important distinction, for it illustrates Clarke's belief that it is the masses of people, and not the leaders, that make and change history. Folk culture and folk tales are made by such people. Clarke and Ellison remind us of these things. O'Meally (1980) focuses on the folk heritage of persons of Afrikan descent as well, giving special attention to Mary Rambo, Jim Trueblood, Peter Wheatstraw, Brother Tarp and Dupre. The folk aspect is extended by Mary Rambo when she sings the blues, which is a form of Afrikan culture. When, at the end of the novel, IM speaks in idioms characteristic of the blues, he is inspired and comforted by it. Similarly, the spirituals and gospel songs sung in the novel celebrate the power that persons of Afrikan descent have summoned to better themselves. O'Meally believes that IM learns that it is his heritage that will free him. Evidences of this culture at play abound in the novel: the sermon on the Blackness of Blackness, IM signifying at Brockway during their fight and playing the dozens[20] to avoid scrutiny while in the factory hospital, again signifying and using the dozens in a meeting with Wrestrum and Brother Jack, and using the dozens again during the uprising, a marked departure from his inability to speak in southern Afrikan American idioms with Peter Wheatstraw. Busby (1991) believes that IM's use of the dozens while in the factory hospital established IM's identity by connecting him to his past. On the other hand, Busby sees Mary Rambo's Sambo bank as a symbol of Afrikan exploitation. IM's inability to get rid of it at that point in the novel was symbolic of his inability to escape his exploitation at that point. Scruggs (1993) sees IM discovering the possibility of remaking Afrikan culture by reshaping what European culture has given to people of Afrikan descent. These critics devalue Afrikan culture and hold it captive to the contributions of an alien, oppressive culture.

Stephens (1999) astutely observes that when IM makes his first speech for the Brotherhood, he inverts the metaphors, talking about collectivism and self-determination. This speech is significant because it expressed IM's views after he had been abused by Bledsoe at the college, by Brockway, and by the doctors at the factory, but before he is indoctrinated into the Brotherhood. At this point, IM is an Afrikan-centered nationalist. Rodgers (1997) sees the novel as IM's quest to transform himself from an individualistic hero into a communal man settling into a place of traditions and shared cultural practices. An example of this concept is Mary's boarding house, which is a southern community in a northern location. As such, it creates a stark contrast with the college and the Brotherhood, which are self-interested institutions.

In an analysis of *Invisible Man*, Tate (1987) looked at IM from a woman's perspective (see also, Bobo, 1995). She sees the succession of women of European descent that IM encounters as a symbolic statement of his growth. The first of these is the forbidden White woman in the battle royal. IM dances with Emma, Brother Jack's mistress, as an equal at the first Brotherhood gathering after he joins the organization. Then there is the wealthy White woman who seduces him while discussing the finer points of "the woman question." And finally there is Sybil, who IM intends to use to gain information about the inner workings of the Brotherhood. Tate's view is that each woman operates on a psychological level as a surrogate mother for IM, and that each delivers him, by degrees, from his false identity with the Brotherhood. Tate viewed the old slave woman in the prologue as an example for IM of the need for action in order for him to obtain his freedom. Hence, his leaving the underground will not suffice; he must commit himself to action. Tate also removes Mary Rambo from the "mammy"[21] tradition to which she appears to be a stereotype. Rather, Tate sees her as IM's surrogate mother. Mary refers to IM as "Jack the Bear" and "John Henry," heroic characters in the folklore of people of Afrikan descent, in an effort to call forth his potential for strength and leadership (Tate, 1987, p. 168). Tate's analysis shows another way in which *Invisible Man* is grounded in Afrikan life and lore. When Mary names IM after heroes, she raises him up, bolsters him. The liaisons with White women, on the other hand, do little other than show how IM fluctuates between clarity of purpose and misdirection. He pursues only one of them, Sybil, and does so only to get a better understanding of the Brotherhood's inner workings. All the other encounters with White women are initiated by them, and show IM as an object of sex, rather than a subject of his own destiny. As such, they are minor incidents in this otherwise densely packed novel.

Larry Neal, a literary critic of Afrikan descent, had varying responses to *Invisible Man* over the years. In his first major critique of Ellison and *Invisible Man,* Neal said that the nationalism and revolutionary aspirations that fueled Richard Wright and his writing, were alien to Ellison. Neal says that writers of Afrikan descent of the 1960s knew who they were, and therefore were not invisible, at least to each other. Neal believed that artists of Afrikan descent had to link their art to the liberation struggles of their people, writing to produce a psychological liberation for them. Dubey (2003) said that Neal and Baraka, editors of the anthology, *Black Fire*, criticized literature by persons of Afrikan descent as the province of the middle class. Afrikan music: blues, jazz, spirituals, gospel, and rhythm and blues, on the other hand, was more representative of a true Afrikan racial ethos because it resonated with people of Afrikan descent from all classes. Therefore these creative forms represented a more collective, as opposed to individualized, psyche. Ellison's work combines literature and forms of music from Afrikan American sources, and in doing so, removes itself from this critique. At one point in his career, Neal came to the conclusion that Ellison saw nothing of value to literature or literary criticism anywhere in African culture (Rowell, 1985). Neal later altered his view. In 1970, Neal did a reappraisal of his earlier position. He recognized that much of the criticism of Ellison came from a Marxian analysis of literature, called "social realism'" (Neal, 1970, p. 31) that did not use Afrikan standards to analyze literary works. Neal saw a strong, spiritually sustaining culture exhibited in *Invisible Man.* Neal rejected his earlier critique of Ellison, seeing that what separated Ellison from the younger writers of Afrikan descent was the issue of activism by these writers. Neal saw *Invisible Man* as a novel operating in the tradition of cultural nationalism as advocated by Karenga (1980) and Cruse (1967). He observed the folkloric and mythic universe operating on the surface of the novel as also operating on the structure of a Louis Armstrong blues. Neal goes on to note that a true Black aesthetic must consist of elements of folk culture: blues, folk narratives, spirituals, gospel, speech and oral history, among others. The myth of the Founder's life as sermonized by Reverend Barbee functions as an object lesson for the young people in the college to follow. For Neal, the Founder's life, too, is folklore. He asserts that such an aesthetic would sift through world knowledge, selecting all that was meaningful and would make persons and communities of Afrikan descent stronger. He concluded that any writer that did not believe that people of Afrikan descent in America had built something powerful and morally sustaining during their 400 years in this country had missed the point. Excellent art, said Neal, is the best propaganda any group can have. Such an art creates another worldview or cosmology that comes out of standards created by that people. Ellison, says Neal, should be admired for keeping Afrikan American culture alive. Neal came to believe that *Invisible Man* was both a public piece of art and a political piece of art. Neal concluded that a writer is a bearer of culture, manipulating the collective conscious and unconscious of their community through their craft.

Baumbach (1986) attempts to place *Invisible Man* within the larger context of the American novel. In so doing, he both misses the place that the novel occupies in the Afrikan American experience, and accidentally makes cogent comments about Ellison's purpose in writing it. Baumbach notes that this is a passage novel, which describes the journey from innocence to experience. By placing the onus for IM's experience on IM, Baumbach shifts blame for his challenges away from external factors, and locates them within the protagonist himself.

Rather than recognize the reality of Afrikan life in America, Baumbach chooses to perceive Afrikan life as an unreality, and to analyze the story as a series of rebirths, rather than as a series of coping strategies tried and discarded when they do not work. IM is present and active in evolving himself, and in the process, is doing what he can to assist his people along the way. His desire to be a leader in service of his people at the college, and in the Brotherhood, was an Afrikan-centered sentiment, even if the ideology and methods he used did not work in the community's best interests all the time. And when he realized that he had been manipulated against his people, he abandoned the Brotherhood and stood with his people in the streets, seeking and finding his place once again.

Mason (1970) also argues that literature written by persons of Afrikan descent in the United States should be evaluated using references, rules, and standards which come from an Afrikan American truth. From this vantage point, Mason views *Invisible Man* as written with a mainstream mentality and in consonance with white literary values. Mason laments that Ellison did not take the opportunity to stand with DuBois in the analysis of history and with Alain Locke in the analysis of art. Each of these authors enunciated Afrikan standards. Mason concludes that Ellison was not trying to eradicate his own invisibility, but rather to make it clear that it existed. Mason accuses Ellison of disposing of his Afrikan will in favor of that of a "Negro" living in a world of white wills. He accuses Ellison of trying to prove the humanity and universality of persons of Afrikan descent, rather than in attempting to solve the problems that being of Afrikan descent in America presents. Gayle (1969) notes that if this was Ellison's quest, it was unsuccessful, as critics and librarians continued to describe and house *Invisible Man* in a section reserved for Negro literature. Mason takes the same perspective that Neal took in his early analysis. Mason misses the grounding of the novel in Afrikan contexts: blues, spirituals, forklore, family, community. He also misses the centrality of the Afrikan world that operates through most of the novel. In short, Mason equates Ellison, who characterized himself as a moderate integrationist politically (Ellison, 1966), with the novel he wrote. The novel differs from the man. Both are Afrikan, but the novel did not have to answer to White publishers, White academics, predominantly White critics and mostly White audiences as Ellison did. Collier (1970) saw the solution to this dilemma for persons of Afrikan descent as the need to assert the emotional certainty of their own humanity. Doing so would render them immune to all the tactics shown in the novel by which persons of European descent intend to subjugate them. IM was seeking this self-confidence throughout the course of the novel. Collier concludes that the message of the novel, and the solution that Mason was looking for, was for persons of Afrikan descent to function between their own individuality, rooted in their racial heritage, and in their status as a member of a nation. In other words, Collier saw the "we" of community as an antidote to the aloneness and isolation of the "I" that resulted in invisibility.

West (1970) looks at *Invisible Man* from a sociopolitical viewpoint, seeing the novel as exploring the issue of being "colored in a white society" head-on in a courageous way (West, 1970, p. 102). He points out that Brother Jack acts as a plantation overseer, manipulating the slaves against each other (IM and Clifton versus Ras) in a deliberate attempt to get rid of Ras and show how the power of the state (the police) was used to exploit the masses of the people in Harlem. Lehan (1970) carries the analysis a step further, giving IM agency and seeing him as active in determining his destiny in the novel. But in focusing on IM's rejection of some Afrikans in the novel, Lehan sees a human issue that goes beyond the boundaries of race, rather than as a series of disagreements as to tactics within a "family" of people of Afrikan descent. Kaiser (1970) says that the novel is profoundly anti-Afrikan, its theme being one of IM against the world. This analysis, while reasonable on the surface, denies the richness of Afrikan culture that is the stuff that breathes life into the novel. IM's every action is to advance himself so that, in his view, he can better serve his people. This is not anti-Afrikan. The attempt to manipulate him by Norton, Brother Jack, and others delays, but does not deter IM. He does not reject Mary. He dislikes aspects of who she is. Similarly, he does not reject Ras or Brockway. Rather, he reacts to their physical assaults on him by defending himself. He did not initiate an attack against Ras or Brockway when these characters did things with which that he did not agree. Rather, he stated his position and attempted compromise until compromise was no longer available. This is in keeping with the Afrikan tradition of compromise and discussion, rather than the categorical rejection and manipulation that are found in characters of European descent such as Emerson and Brother Jack, as well as in their surrogates, such as Dr. Bledsoe.

This view is important, given the time of the novel's publication. Afrikan American men had been participants in World War II, equal in the right to fight and die to save America and democracy from dictatorships abroad, but were still denied equal citizenship at home (Franklin & Moss, 2000). At that time, the issue of race (as well as those of class and gender) was put aside by men of Afrikan descent in the interest of fighting for democratic ideals and against Communist and Nazi alternatives, while inequality continued to reign in the United States. Thus, to write, and have published, a novel that confronts those issues so directly was indeed a forward and daring step. It signaled a self-consciousness among Afrikan American authors that paralleled the innovation and expansion of blues and jazz, the growth of a confident Afrikan American scholarship, and the movement toward civil rights and Black Power, that were always there, and were to flower in the 1950s and 1960s. The challenge with West's interpretation is that *Invisible Man* is not *only* a novel about being of Afrikan descent in a society dominated by people of European descent. Rather, it is simultaneously about Afrikan society, about European society, about the spaces where these two societies overlap and intersect, and the connections made between them as members of each society interact with each other. Ellison has painted a picture of two worlds: one Afrikan and one European, in dynamic interaction with each other. In fact, the Afrikan world in the novel is more complete, more well-defined, and contains more detail than the European world. These worlds interact on the bases of shared space, shared history, shared destiny, and shared interests. They interact as equals in that each world has its own cultures, mores, spiritual systems, and governing structures. To see the Afrikan world in the novel as anything less than complete is to miss the totality of what Ellison presents, and to minimize the import of the Afrikan world in the novel.

Moses (1989) places Ellison squarely in the swaying political and social winds of Harlem in the 1930s and 1940s, where the Garvey and Rastafari movements that are mirrored in the novel were centered. This is important because of Ellison's own denial of Afrikan American literary influences on his writings (Ellison, 1966). Savery (1989) affirms this, noting that Ellison's ". . . existentialism rejects the totally negative vision of the world, the lack of consciousness, transcendence, or social responsibility . . . . What Ellison accepts is the world of possibility, consciousness, and struggle; the world of ambiguity over the world of cold, predictable logic . . ." (Savery, 1989, pp. 66-67). Savery's comparison of *Invisible Man* to the music of the blues, with its theme of emerging selfhood and transcendence of pain, is both constructive and useful. However, when Savery says that IM encounters others who know who they are, whereas he is a creation of others rather than of himself, Savery misses the point. IM exhibits a constantly emerging self and recognizes the struggles that the self endures in an alien land. By pointing to the positive characters in the novel, such as Peter Wheatstraw, the man selling the yams, Mary Rambo, and IM's grandfather, for example, the author and his character recognize that keys to survival and prosperity do exist within Afrikan American communities. Further, Savery realizes that those personality types can yield a healthy person, whether one views the landscape as oppressive or liberating. These characters are everyday people of Afrikan descent, heroic in their love of life, family, community and culture.

Wilber (1999) makes the provocative claim that some thirteen genres of fiction are missing from the lexicon of Afrikan American literature because they are not being written, not being published, are not available, or for a combination of those reasons. As a result, what is published by mainstream publishers, virtually the only publishers available to Ellison at that time that he wrote *Invisible Man*, reflects the negative stereotyping of the publishers, rather than that of the writer/ creator. With this understanding, is it not also true that, as a writer, Ellison was kept running? Did he not have to couch his writing so that it could and would make it into print? And therefore, is there a deeper level available to the reader of *Invisible Man* that goes beyond the stereotypes, just as in Afrikan folktales, where the message is hidden in symbols and metaphors, such as in the Brer Rabbit tales?

Vogler (1974) takes this point regarding Ellison and applies it to IM. He says that just as the person of Afrikan descent turns their everyday struggles in American society into art, IM turned his struggles into this novel, by writing down the challenges of his past. More importantly, he see IM's quest as one for a group in which he can find freedom and his identity. This view is clearly biased against the rich Afrikan culture in which IM already operates, and from which he springs. IM does not need another group, such as the Brotherhood, to find himself. He has a group; he needs to return to it and reorient himself within it.

Howe (1974) believed that Ellison could not have written with the distance from his topic that was demanded of writers of the time, because to do so was both morally and psychologically impossible for Ellison as a person of Afrikan descent. Hyman (1974) saw Ellison's creation of the novel as an exercise in freedom, in the same way that a jazz musician learns how to play their instrument before they are free to improvise on it. Strout (1989) sees Ellison's personal vision in his writing, as revealed in Ellison's own analysis of the book. Strout says that for Ellison, knowing who one is, is a precondition for freedom, but the discovery of who one is, is both a communal and individual process. From this viewpoint, being and becoming do not occur in isolation, or in linear fashion. Rather, they are diunital, occurring at the same time. IM's identity is always present, and evolves as the novel progresses. He acts, makes errors in judgment, recovers, and acts again. This is the process of growth in action.

Kent (1974) points to the multiple levels at which some of the symbols in the novel operate to create a rich and deep sense of Afrikan

American culture. The chain link from the era of slavery in the hands of Dr. Bledsoe and Brother Tarp's link from the chain gang are connected, expressing oppression and survival. Mary Rambo's Sambo bank and the dolls controlled by Todd Clifton resemble Dr. Bledsoe, controlling others such as IM, and being controlled by others, such as the trustees. The blues is used to evoke emotions regarding the places and situations in which IM finds himself. And the black substance in the white paint at the Liberty Paint Factory symbolizes the Afrikan labor that built the wealth of the United States. Kent quotes Ellison on Black folklore as "announcing the Negro's willingness to trust his own experience, his own sensibilities as to the definition of reality, rather than to allow his masters to define these crucial matters for him" (Kent, 1974, p. 162). In an earlier work, Kent (1972) says that Ellison used folklore and cultural traditions to evoke a deeper knowledge of Afrikanity. Despite the analysis of many critics, who force Ellison into an assimilationist tradition in their analyses, Ellison's position points to a selfhood and a race-based and culture-based concept of reality that springs from within the Afrikan American experience, rather than being imposed from without. Thus while the characters that Ellison created and the symbols he used may be common to the American experience, they have particular meanings to persons of Afrikan descent, and are of particular value in determining psychological groundedness and health for people of Afrikan descent.

Forrest (2000), himself a novelist of Afrikan descent, says that the grandfather, as the oldest ancestor within IM's family's memory, passes on the family's oral history and wisdom as griots and griottes[22] did in Afrika. Forrest sees IM's recollections of his personal history and his family history as a second ancestral voice. And he sees the symbols, such as Tarp's chain link and the Sambo dolls, as part of a racial memory and race consciousness that connect him with and make him responsible to an ancestral community. Like all cultural wisdom, some of it springs from trials, and some comes from triumphs. *Invisible Man* gives us Brother Tarp's humanity and Dr. Bledsoe's greed. Both examples provide life lessons. One illustrates how to be Afrikan at the highest level, while the other illustrates the allure of greed and power. Both men arose from humble beginnings. Both had been enslaved. Tarp was poor but free. Bledsoe was wealthy, but still enslaved by his own ambition. These lessons offer pathways for Afrikan male identity development.

Tate (1998) utilized psychoanalytic theory to analyze selected Afrikan American texts. Predicting responses from the Afrikan American community for using mainstream psychology in her analyses, she noted:

> I am well aware that there are many who would contend that the imposition of psychoanalytic theory on African American literature advances Western hegemony over the cultural production of black Americans, indeed over black subjectivity. This is a curious defense in light of the fact that black literary criticism often regards black characters as signifiers of racial ideologies rather than as subjects, even as it focuses on the psychological trauma of racism. (Tate, 1998, p. 192, n. 6)

Tate's statement goes to the heart of this discussion. First, mainstream psychoanalytic theory *is* a hegemonic imposition on Afrikan American life and culture. It *is* an externally imposed system of analysis based on values that place people of Afrikan descent at the margins of existence, rather than at its center. Second, to state that this is true is not a defensive response. Rather it is an initial salvo in a war to reclaim critical social and cultural space from the external control of White intellectuals and those who use their formulations to perform analyses of Afrikan subjects (Carruthers, 1999). This book removes the psychology of Afrikan male identity development from the domain of European power and control. Third, the use of Black characters as "signifiers of racial ideologies rather than as subjects" is precisely the stance that a conscious person of Afrikan descent would take in moving from the psychological trauma of racism to the redefining of self in Afrikan American psychological terms.

Conclusion

Invisible Man offers an impressive example of a novel that can be used to develop, evaluate and clarify the theoretical models of Afrikan male identity development evolved by Afrikan-centered psychologists and create a new model. Unlike recent theorists of European descent, who are investigating the centrality of feelings to identity (de Quincey, 2005), people of Afrikan descent have long recognized that feelings, thoughts, spirituality, and communality all play an important part in the identity of people of Afrikan descent. That discussion of identity development is the focus of the next chapter.

CHAPTER IV

Discussion, Theory-building, and Implementation

Practice without thought is blind; thought without practice is empty (Nkrumah, 1970, p. 78).

The question of *identity* lies at the heart of the African quest for global agency and liberation (Carr, 1997, p. 287, [*italics in original*]).

Introduction

The purpose for this treatise is to answer a series of questions. Is *Invisible Man* an Afrikan-centered novel? Is IM an Afrikan-centered human being? Are there features of what IM went through in his life that can be culled out and fit into an Afrikan-centered theory on Afrikan-centered male identity development? And, if there are pieces missing from what the Afrikan-centered theorists have offered which are found in our protagonist's trek through the novel, can those issues be pulled out and add to existing theories to create a coherent theory of identity development?

The novel *Invisible Man* invites the reader to reflect on his or her own identity, to put it in focus, and perhaps, in order. The critics of *Invisible Man* and of Ellison raised many issues concerning identity, both of the character of IM in the novel, and of the author of the novel. These issues are summarized below, utilizing the Black Aesthetics framework for determining an art that is both artistically creative and socially relevant. Such an art is a precondition for utilizing its characters in contributing to the theoretics of identity development from an African-centered perspective.

Does *Invisible Man* reflect an Afrikan culture in America? Clearly it does. But before delving into the novel again, and extracting from it the identity framework that it contains, a review of the elements of African-centered psychological theory that might inform an emergent theory of identity is in order.

Afrikan Psychology and Identity Development: A Review of Theories and Concepts

In the 1960s, Ellison was vilified by a younger generation of authors and critics of Afrikan descent for his political views, which were anything but nationalist. However, Ellison consistently argued that the culture of people of Afrikan descent was adaptive and vibrant. He said of Afrikan culture in America: ". . . there is also much of great value, of richness, which, because it has been secreted by living and has made their lives more meaningful, Negroes will not willingly disregard" (Ellison, 1966, p. 302). And in a 1967 edition of Harper's Magazine, Ellison defended the Afrikan culture in Harlem from those scholars who viewed people of Afrikan descent only as oppressed:

> I don't deny that these sociological formulas are drawn from life. But I do deny that they define the complexity of Harlem. They only abstract it and reduce it to proportions which the sociologists can manage. I simply don't recognize Harlem in them. And I certainly don't recognize the people of Harlem whom I know. Which is by no means to deny the ruggedness of life there, nor the hardship, the poverty, the sordidness, the filth. But there is something else in Harlem, something subjective, willful, and complexly and compellingly human. It is that something else' that challenges the sociologists who ignore it, and the society which would deny its existence. It

is that something else' which makes for our strength, which makes for our endurance and our promise. (Thomas & Sillen, 1993, p. 46)

The "something else" to which Ellison alludes but does not define is made up of the collective Afrikan unconscious core identity and the behavioral manifestations of it in Afrikan-centered life activities. It is to that something else that I AM appears to be headed when he leaves his hibernation in his underground abode.

Asante (1980) spoke of a universal Afrikan consciousness as a connection between the pasts, presents, and futures of cultures and peoples of Afrikan descent. Bynum (1999) expanded on Asante's theory, noting that the human and cultural roots of all human beings stem from Afrika, and that a collective Afrikan unconscious exists that animates all Afrikan peoples (as well as all other members of the human family). Bynum says that the Afrikan unconscious consists of rhythm, spirit, bioenergy, community consciousness, and both intrapsychic and interpersonal principles. King (1994) called this collective unconscious or Afrikan essence "the Black Dot." Diop (1991) also gave definition to an Afrikan sensibility, which he called the Afrikan cradle of civilization. According to Diop, this sensibility is matrilineal, monotheistic, communal, collective, congenial, non-racial in social relations, and complementary on all other levels. Lee and Bailey (1998a) define this Afrikan sensibility as consisting of kinship, cooperation, mutual respect, commitment, and spirituality.

Other theorists come to similar conclusions. Herskovits (1958) asserted that people of Afrikan descent brought their cultures with them when they were stolen from Afrika and arrived in the West. DuBois

(1961) made the same point earlier in the century, saying that Afrikans in America had a "double consciousness" (p. 3), one Afrikan and one American. DuBois's point is that Afrikans in America did not lose their Afrikan-ness in the Americanization process. Boykin (1979) asserts that the cultural strengths and adaptive skills that persons of Afrikan descent have exhibited in order to survive the challenges presented to them by American society are evidence of inherent resilience and survival skills. Jones (1991) agrees that a form of Black psychology has been practiced by persons of Afrikan descent in this country since the first enslaved Afrikan person arrived in America. Jones' TRIOS theory (1991), which consists of the elements of time, rhythm, improvisation, oral expression, and spirituality, accounts for and explains the psychology practiced by persons of Afrikan descent.

Kambon (1992) discussed an Afrikan sensibility, which he called an Afrikan worldview. This worldview consists of kinship, community, common experience, symbolism, and spiritual association. Dixon (1976) also discussed an Afrikan worldview, in which man is part of or one with nature. With these concepts as background, one can see that persons of Afrikan descent combine thinking and feeling as a method for knowing. The past, present, and future are part of a continuum, rather than being discrete entities unto themselves. Nobles & Goddard (1993) proposed an eight-part theory of Afrikan culture. In it, they offer the concept of consubstantiation. Consubstantiation means that all elements (humans, animals, inanimate objects) are created of the same substance. Interdependence means that all elements in the universe are connected. Egalitarianism means that all relations are harmonious and balanced. Collectivism consists of codes of conduct based on the idea of group and/or collective survival/ advancement. Transformation as change is movement toward a higher level of functioning. Cooperation means that things function based on mutual respect and viability. Humaneness means that behavior is governed by a sense of vitalism and viability. And synergism is the notion that the sum of complementary actions is greater than the total effort of individuals (Nobles & Goddard, 1993).

Nobles (1972, 1976, 1980) offered an understanding of the difference between Afrikan and European concepts of psychology with his explanation of the extended self. Implied in this concept is that life for persons of Afrikan descent is communal; each such person is defined by and responsible to the group for their existence. Such a community is continuous, with elders, adults, and children joined by ancestors and those yet to be born, reflecting Afrikan mystical traditions (Tetteh, 1997). Akbar (1976) also emphasized the primacy of the group for persons of Afrikan descent. Azibo (1988) proposed that the Afrikan personality exists in three parts: an inner core, an outer core, and an action component. The inner core is biogenetically determined and unchangeable. It consists of spirituality, energy, rhythm, and a self that is both personally defined and collectively determined. The outer core contains values, beliefs, attitudes and opinions that form a belief system and produce behavior in concert with the inner core (Azibo, 1988). Kambon (1992) expanded on the concept of the Afrikan personality. He agreed with Azibo that it is biogenetic, traits inherited from one's ancestors and which are unalterable by external influences. With this in mind, Kambon discusses two aspects of the Afrikan personality: the African Self-Extension Orientation or ASEO, and the African Self-Conscious, or ASC. The ASEO is the biogenetic, deep-rooted energy that exists in each person of Afrikan descent and extends to others of Afrikan descent. It is a storehouse of ancestral Afrikan knowledge, a concept similar to Azibo's inner core. The ASC is the conscious expression of the ASEO, similar to Azibo's outer core and action component. When a person of Afrikan descent is in a fully nurturing environment, the ASC and the ASEO become one. Thus, Afrikan-centered theorists agree that a biogenetic, immutable Afrikan sensibility exists in all persons of Afrikan descent everywhere in the world. To this core self is added an outer core (an Afrikan worldview) and a behavior component which, when properly nourished, acts in concert with the inner core.

Black Identity Theories

Using Karenga's (2002) framework, current reformist psychological identity theories, which can also be called Black psychology identity theories, that focus on persons of Afrikan descent, contain little of the information and perspectives discussed above, and lack an Afrikan worldview. Cross's Nigrescence theory (1979), as well as those authored by Thomas (1971), and Milliones (Cross, 1980), are similar to the elaborations of Cross's theory by Parham (White & Parham, 1980) in that they assume no Afrikan unconscious and no Afrikan worldview. In short, from the perspectives of these theories, Afrikan people are not born with a connection with their Afrikan ancestors, and are not part of an Afrikan continuum. Cross's pre-encounter stage assumes that persons of Afrikan descent are born with a Eurocentric orientation that rejects Afrikan culture and connection in favor of European culture, thought, and truths (Hall, et al., 1972). Watts, et al. (1999) offered a similar theory, but called their theory "sociopolitical development" rather than identity development. In the first stage of this model, the person of Afrikan descent is oblivious, both to their own Afrikanity and to their oppression in the United States. As a person progresses through the stages of this model, they become more sociopolitically aware, and end up a part of a process that works toward liberation. Here, again, people of Afrikan descent are at the periphery; European hegemony as practiced in the United States is at the center of the analysis and discussion. Azibo (1988) critiques all of these theories for being part of what he called the negativist school because they do not start or center their analysis with people of Afrikan descent, but instead adopt an alien theoretic system.

They then construct the rest of their theories by observing how people of Afrikan descent respond within that system. Azibo criticizes the Nigrescence theorists for failing to see that persons of Afrikan descent have an innate drive to survive, which Kambon (1998) calls a "survival thrust" (p. 120), that should be the focus of the first stage of their theories. Azibo also criticizes these theorists for the final stage of their theories, where persons of Afrikan descent are presumed to have moved beyond race to focus on the larger environment of the United States. Azibo and Robinson (2004) refer to this group of Black identity development theories as transformational theories. They continue to see the Cross, Thomas and similar models as flawed, since they all begin by viewing identity development as being independent of an Afrikan inner core. In other words, if the theory does not have as its basis and focus the survival and prosperity of Afrikan peoples, that theory is flawed, and persons who fit those theories are not normal, in the Afrikan sense of normality. Azibo and Robinson conclude that the Cross model and its progeny began with a flawed design. Subsequent studies whose goals are to validate the original theory also validate the error made in the theory.

Afrikan-centered Identity Theories

Semaj (1981) offered a three-stage theory of Afrikan identity that begins with a person who had an Afrikan identity which they lost or replaced with an individualistic materialistic, Eurocentrically based self. The person then reconnects with their Afrikan identity in stage two, but still perceives themselves as a victim of anti-Afrikan supremacy, unable to alter their lot. In the third stage, a shift occurs, and the person re-centers themselves in a collective extended identity, focusing on the survival and prosperity of persons of Afrikan descent. Williams' (1981) three-part theory, known as WEUSI, as previously discussed, contains an Afrikan core, or "WE," that is the same genetic, cultural, psychological, and spiritual core as was expressed in the theories of Akbar (2004), Asante (1980), Bynum (1999), Diop (1974, 1990), Dixon (1976), Kambon (1998), and Nobles (1986). The family, community, national and worldwide collectivity of Afrikan peoples is called "US" in Williams' theory (1981), and the "I" refers to the expression of these collective Afrikan traits in each individual, and includes spirituality and rhythm.

Semaj (1981) and Williams (1981) have both created Afrikan-centered theories. Absent are the paradigms and frameworks of Europeans imposed on Afrika and Afrikan peoples. Rather, the survival of Afrikan people and realization of the fullness of Afrikan existence are the constants of these theories. However, neither theory discusses development: rather, they assume that Afrikan identity is a birthright and a constant. There are those in the Afrikan community in the United States who are in need of identity development. For them, these theories are useful, but not sufficient.

Nobles & Goddard (1996) list four behaviors that grow out of the eight cultural precepts they had previously enunciated and which are discussed above (Nobles & Goddard, 1993). Authentic behavior is behavior that is consistent with Afrikan culture. Adapted behavior is consistent with the culture but has been modified in response to the environment. Adopted behavior borrows behavior from another culture without analysis or selectivity. Aberrant behavior is opposed to the culture and has the potential to destroy that culture's fabric (Nobles & Goddard, 1996, p. 13). These cultural precepts and behaviors are useful in implementing a theory of identity development. In their work on Afrikan American high risk youth, Nobles & Goddard (1993) proposed methods to implement the information taught to the adolescents so that they absorbed and were able to act upon it. Nobles and Goddard suggested six modes of presentation of the material: dramatic consciousness (analogizing the application of information to a drama in which the young person is an active participant); mind modeling (using the lives, challenges and triumphs of other Afrikan people to illustrate the path to overcoming similar difficulties); image and interest discussion/dialoguing (the ability to find the Afrikan-centered position in any discussion and respond appropriately); culturally consistent problem-solving (developing solutions that will lead to progress for the individual and reinforce those solutions); metaphoric memory (going from the known to the unknown by comparing the known with the unknown, through the use, for example, of proverbs, lyrics, and stories); and analogical thinking (using a reference system to notice similarities between events and experiences); (Nobles & Goddard, 1996, pp. 123-124). The result is the involvement of the young person's thoughts, feelings and actions in moving toward the desired goal. This framework would fit well in operationalizing a model for working out those aspects that a person of Afrikan descent requires in order to re-center themselves in Afrikan-centered reality.

Sutherland (1993) proposed a four-part model of identity of individuals of Afrikan descent, using the work of Afrikan-centered psychologists. She labels three of the parts as "nonideal orientations" and one part an "ideal orientation." The nonideal orientations are reactive operations of persons victimized by oppression. In the ideal orientation, "the authentic struggler" refuses to adjust to oppression (Sutherland, 1993). However, no theory was offered and no platform was proposed for the transition from the nonideal to the ideal orientation other than to immerse the person in the literature of those who have described how others have deviated from their Afrikan identity, such as Fanon (1967a) and Asante (1980).

Of all the theorists studied, only Akbar (1991) discussed Afrikan male identity, which he said consisted of self-respect, self-knowledge, and self-definition. Akbar (1994) also discussed an aspect of the soul called Seb taken from the Kemites, which is the self-creative power of being. As previously discussed, Akbar (2004) enumerates four disorders that occur when persons of Afrikan descent become, in Kambon's terms, psychologically and/or culturally misoriented (Kambon, 1992, pp. 136-138). Three of the four disorders enunciated by Akbar are the result of a person of Afrikan descent rejecting an Afrikan worldview in favor of the actions, thoughts, and beliefs of Eurocentric society (Akbar, 2004). Those who have alien-self disorder, where one fully embraces a European worldview, anti-self disorder, where one is hostile to anyone who expresses an Afrikan worldview, and self-destructive disorder, where persons engage in individualistic negative behavior to survive oppression, reject or are disconnected from Afrikan culture and Afrikan peoples. The fourth set of disorders, organic disorders, occur as a result of societal oppression being internalized, resulting in maladies such as malnutrition, incarceration, or substance use. Akbar says that in all of these disorders, the innate rhythm of Afrikan people has been disturbed or disrupted in these individuals, either physically, mentally, and/or spiritually, and must be restored for healing to occur. Kambon's (1992) prescription for the healing of psychological or cultural misorientation consists of: re-education toward a Pan-Afrikan, nationalist consciousness and the practice of an Afrikan-centered study of Afrikan history and the universe; elimination of non-Afrikan names and adoption of Afrikan names; removal of Eurocentric symbols from the home and other places where persons of Afrikan descent gather, and replacement with Afrikan rituals, structures, and symbols. Lee and Bailey (1998b) suggest that providing Afrikan males with role models to consult and emulate will be a key ingredient in their healing process.

Afrikan-centered identity development theories differ markedly from the other identity theories discussed in that they begin with an Afrikan-centered focus. That is, in these theories, an Afrikan child enters existence with an Afrikan unconscious, a biogenetic core that resonates with the continuity of Afrikan life, and vibrates with the rhythm of Afrikan peoples, the collective Afrikan unconscious. This essence is never lost, but can become temporarily sublimated to the surface challenges that living in a Eurocentric environment evoke, such as set of beliefs that are contrary to the essence of who Afrikan people are. Baruti (2005) warns that identity development in the Afrikan-centered sense must be inculcated in Afrikan children while they are still under their parents' wings. The goal of an Afrikan-centered identity theory is, then, to restore that Afrikan essence to the central place in the life of each person of Afrikan descent. As for Afrikan males, their role as part of the balance of life (Fu-Kiau, 1991, p. 66) with females of Afrikan descent must also be considered in advancing any theory of Afrikan male identity development.

With these ideas and concepts in mind, we revisit *Invisible Man* to see what components of a theory of Afrikan-centered male identity development can be found in the novel that are in consonance with these Afrikan-centered theories.

Invisible Man and Afrikan-centered Theories

Collective Afrikan unconscious. The first question to be asked and answered is whether IM has the innate, biogenetic sensibility to which Asante (1980), Bynum (1999), Kambon (1998), and Azibo (1996b) refer. The answer is that he does because IM is of Afrikan descent, but that we see little direct evidence of it in the novel. Bynum (1999) speaks of community consciousness, a consciousness of being part of and one with a community of people of Afrikan descent. IM showed this sense of community at the college. He was remorseful that he had to leave, believing that he had put the community at risk by exposing the trustee, Mr. Norton, to the Trueblood family and the Golden Day, less desirable aspects of life at the college. This was important, since IM considered the college as synonymous with his identity. He again showed a community consciousness when he stood with the Afrikan community in Harlem against the eviction of the elderly couple, and when he stood with the people of Harlem as he gave his first speech for the Brotherhood. His joining the Brotherhood was also an attempt, in his mind, to be of service to the Afrikan community. Indication of this consciousness is also exhibited during his stay at the factory hospital. After he is electroshocked, he cannot remember his own name or the name of his mother, but he does remember an Afrikan folktale, and uses its wisdom to outsmart the White doctors who are assaulting him. This is not, however, evidence that IM possesses a core Afrikan consciousness.

IM also evidences a connection with and some understanding of the communal, continuous nature of the Afrikan community. He sporadically exhibits what Bynum (1999), Kambon (1998), and Akbar (2004) call the primacy of the group, and what Nobles called an extended self. He takes responsibility for the negative effects his mistakes could have on his community. Those mistakes include the verbal error when he gave a speech at the battle royal, the Norton fiasco, and his sense, near the end of the novel, that he was actively participating in the manipulation of his people by the Brotherhood. He also takes responsibility in positive ways. He speaks for the elderly couple who are being evicted from their home. While doing so, he recalls how hard his mother had worked to raise her family, and equated his family to this elderly couple, realizing that their fate and his were intertwined. He stands up for his people and against the Brotherhood at the end of the novel, signifying his connection to and with people of Afrikan descent in Harlem. But the question of IM possessing an Afrikan unconsciousness still remains unproven.

IM's journey also includes communication with ancestors and elders, who are part of the continuous Afrikan community as defined by Bynum (1999), Akbar (2004), and Nobles (1986). IM is visited by memories of his ancestors in the form of dreams about his grandfather and his mother. IM was respectful toward Dr. Bledsoe, both because IM admired Bledsoe and wanted to follow in his footsteps as a leader of his race, and because he was an elder. He is also respectful toward

Brockway because he is an elder. IM only hits Brockway in self-defense after Brockway hits him. He accepts advice from the vet and from Mary Rambo, both elders in his world. He seeks advice from Brother Tarp, also an elder. He learns from Peter Wheatstraw and from the man selling the yams, both of whom are elders imparting knowledge to him. And he seeks counsel with Tod Clifton, someone of his own age group, showing the continuity of the connection across the generations. But even these actions do not prove the existence of an Afrikan consciousness in IM.

Azibo (1988) describes a behavioral or action component (the ASC in Kambon's framework) that operates with the other component(s) of the Afrikan unconscious. Through much of the novel, IM acts, often believing that he is assisting people of Afrikan descent. However, his actions are not, strictly speaking, Afrikan-centered. That is, his actions do not have Afrikan freedom as their sole focus. Rather, IM is often swayed by Eurocentric ideas, such as those propounded by Bledsoe at the college. IM wanted to emulate Bledsoe (and Booker T. Washington) and become a leader of his race. However, he had yet to discern what that meant. Similarly, IM is wooed by the possibility of a job with the Brotherhood, and ends up being manipulated by them into creating the preconditions for a civil unrest that will victimize the Afrikan population of Harlem. He allows the Brotherhood to convince him to change his family-given name, and moves out of Harlem into a White part of New York. The clearest indication that IM was disconnected with the Afrikan community in Harlem was when he concluded that those of Afrikan descent, such as Mary Rambo, thought in terms of "we" whereas he thought in terms of himself, "me" (Ellison, 1947, p. 316). Hence, if IM does possess that inner core Afrikan unconscious, he shows little evidence of possessing the outer core and behavior, using Azibo's model, or the ASC, using Kambon's model, that activates the inner consciousness.

But IM also acts in ways that redeem himself and Afrikan people. He refuses to lie to White people as Bledsoe suggested (though he does lie to his family about his move from the college to Harlem). In his first speech for the Brotherhood, he inspires hope in the people of Afrikan descent in Harlem by allying himself with them and their plight. And near the end of the novel, he joins in with a group of Harlem residents during the civil unrest, forsaking his leadership role, and becoming a part of the larger family of Harlem. He realizes that they can act of their own accord; they do not need him to lead them. At the end of the novel, IM is about to re-emerge. He enunciates the commitment to a social role with his people. As such, his seclusion has resulted in him acquiring, or re-acquiring an Afrikan-centered strategy for action (Baruti, 2005). Nonetheless, IM's actions are largely self-centered: often, he subverts his innate Afrikan consciousness so that he can get ahead as an individual. IM does possess an Afrikan-centered consciousness that is struggling to balance itself within him. His unconscious is part of the collective unconscious of Afrikan people (Bynum, 1999).

Kinship and family. For Bynum (1999), Kambon (1998), Dixon (1976), Nobles (1986), Akbar (2004), and Azibo (1988) alike, kinship is an important part of the Afrikan unconscious. IM may have been connected to family and community prior to going to the college, but he starts to lose the connection with them as he attempts to prove himself worthy at the college. He occasionally wrote letters to his family throughout the novel, but repeatedly chooses not to communicate directly with them and go to see them when they could have supported him. The first of these incidents comes when Bledsoe dismisses him from the college, and he chooses not to go home and to the support of his family, because he is concerned that they will be disappointed in him. On the bus heading to Harlem, IM again encounters the vet, who acts as a surrogate father. The vet suggests that IM needs to be his own family by finding himself (Ellison, 1947). When he arrives in Harlem, he writes to his family to tell them that he is doing well, when in fact is has no job and is not optimistic about his prospects. In so doing, IM appears to be motivated to make his family proud of him, and to save them from worrying about him and his situation. IM moves out of a hotel an into Mary Rambo's house, where Mary nurses him back to health after the factory hospital incident, and encourages him with her supportive words. He eventually moves away from Mary's house and from Harlem when he goes to work for the Brotherhood and they isolate him from the Harlem community by selecting where he would live. He also accedes to the demand made by the Brotherhood that he not communicate with his family. When IM accepts the new name that the Brotherhood gives him, he denies and discards his real family name. In his first Harlem speech for the Brotherhood, he tells the crowd that he has found his family, which is ironic, because he has moved away, physically and ideologically, from Harlem and from his family of origin. Brother Tarp and Tod Clifton become a sort of substitute family for IM, but since both are also members of the Brotherhood, they are subject to the whims of that organization's White leadership. IM has had no friends during the course of the novel (unless Tod Clifton could be considered a friend), has very few genuine relationships, such as the one he has with Mary Rambo, and has temporarily chosen the isolation of the underground to life as part of the Harlem community. As he ponders re-emerging from his underground existence at the end of the novel, he thinks briefly about returning to the South, his home, but rejects that idea. Hence, IM appears to have lost, or severed his connection with his kinship network, his support system. In short, IM does not have an Afrikan-centered connection with family and friends. He also has rejected the family and friends of the Brotherhood, his adopted family, as they have rejected him. His extended self has been disrupted.

Spiritual connection. IM also is strikingly devoid of spiritual connection. He spends time in the chapel at the college, but this appears to be a religious and not a spiritual duty. While in the chapel, his mind and heart are on his impending fate rather than on spiritual issues. He listens to Reverend Barbee's sermon about the college's Founder, but gains no upliftment from it. Similarly, in the prologue, he dreams of a sermon on the "Blackness of Blackness" (Ellison, 1947, p. 9), but again he draws no sustenance from it. This dream evolves into the image of an old slave woman singing a spiritual. But again, IM is interested in the woman, not the song. At no time during the course of the novel does he pray or meditate. And at no time during the novel does he seek the support of or connection with a higher force or source, especially when he is in need, such as when he is in the factory hospital, or as he is struggling to find employment when he arrives in Harlem.

Transubstantiation. It is a sense of hope that IM inspires in the reader at the end of the novel, a hope that Nobles and Goddard (1993) called transubstantiation, the movement toward a higher level of functioning. Early in the novel, IM laments that he has to leave the college, which he saw as the place that framed his entire identity. In this regard, the vet was acting as a true elder when he urged IM to find himself by being his own family if he were going to move to a place where he had no family. As IM leaves the factory hospital, he comes to the conclusion that when he understood who he was, he would be free. In fact, he would not be free of European oppression, but would be reconnected to/with his Afrikan family, and thus freed of having to make decisions without assistance from those he trusts. But this does not occur at that point. As he recuperates at Mary's home, he wonders who he is. It is shortly thereafter that he encounters the man selling yams, which reminds him of home (and who he is). Soon he gets confused again by adopting the name and identity given to him by the Brotherhood.

But gradually he reclaims himself. He realizes that others wish to sacrifice him for their own good (Ellison, 1947, p. 505). At that point, he also recognizes that he has a connection with the Afrikan people of Harlem. He decides to follow his grandfather's advice, and "yes" the Brotherhood to death, trying to look out for the best interests of the people of Harlem and getting back at the Brotherhood who he believes has betrayed him. In so doing, he again begins to reshape his identity. IM has found himself at the end of the novel. He knows who he is: he is invisible to the white world, and partially visible in the world of Afrikan people. He has emerged from the dream of the slave woman who birthed four sons by the slave master, and kills her master to gain her own freedom. He feels a deep connection with this woman, and with the blues that he hears before and after the dream. He is connected with an understanding of the quest for survival, both his own and that of his people, an aspect of Kambon's (1998) and Azibo's (1988) Afrikan unconsciousness. He is honest with himself, recognizing that some of his problems were of his own making, while others were difficulties he encountered in the world around him. His commitment at the end of the novel is to himself first and then to his people. He knows who he is, knows that he is of Afrikan descent, and accepts his responsibility toward his people.

Identity Development Theories and Invisible Man

Nobles (1998) notes that "the problem with Black Identity theory is that it represents only a limited (albeit damaged) aspect of what it means to be African (or not to Be)'" (p. 189), because the theories operate on a European worldview and paradigm that negates Afrikan genius and culture. Myers and Haggins (1998) point out that persons of Afrikan descent stuck in a suboptimal system of values and practices find themselves struggling to make sense of their reality. The pre-encounter stage of the Cross school of identity development theories assumes no Afrikan core or unconscious, and the final stage of these theories finds the person of Afrikan descent focusing on problems common to all of humanity, and not on the need to continue the flowering of persons of Afrikan descent in the United States (Taylor, 1998). Therefore, they are not useful in determining identity development from an Afrikan-centered perspective. In addition, IM does not "recycle" through stages, as proposed by Parham (Cross, Parham, & Helms, 1991; White & Parham, 1990). The worldview frameworks added onto the Cross/ Thomas theories by Helms (1993) and by Cross, et al. (1991) do not alter this result. Rather IM appears to be partially conscious, alternately swayed toward and away from a more Afrikan-centered consciousness by the influence that the Eurocentric elements around him have on him. To get an Afrikan-centered perspective, the theories of Semaj (1981), Williams (1981), and Akbar (2004) must be used, with the goal being the higher level of functioning in the best interests of people of Afrikan descent that Goddard and Nobles (1993) enunciated in their concept of transubstantiation.

Karenga (1980) notes that people of Afrikan descent in America lack a vocation, a collective vision and mission to build or rebuild the Afrikan nation in America. One aspect of this challenge is what he calls an identity crisis - we can have no national vocation if we don't know to what nation we belong or what to call ourselves. Moreover, in Karenga's view, if we don't see ourselves as a nation, a people who share a common history, common culture, common life-conditions and chances, and a common self-consciousness, it would be impossible to establish a collective vocation (Karenga, 1980, p. 104). Wallace (2002), concurs, stating that identity in its Afrikan-centered and therefore socially conscious context, must be grounded in Afrikan and Afrikan American realities.

Semaj's theory makes sense. It begins with the person of Afrikan descent possessing an Afrikan identity, similar to Kambon's ASEO and ASC and Azibo's Afrikan inner and outer core, an identity which gets temporarily displaced by a Eurocentrically based self. IM comes from an Afrikan family, but we do not know whether his initial values were Afrikan-centered. His family wanted him to be educated, or they would not have allowed him to accept the scholarship to the college. His family may not have known that by going to the college, IM was going to be trained rather than educated, trained to work in service of White people rather than educated to work in service of people of Afrikan descent (Woodson, 1933). When IM goes through the trials in New York, initially unable to get a job, and then being injured at the Optic Paints factory and again in the hospital, he is Semaj's second stage. He is still operating as a victim of anti-Afrikan supremacy and unable to assist himself or his people. When he joins the Brotherhood, he believes he has found a way to assist his people, but eventually he learns that the Brotherhood has the interests of its White leadership in mind when it acts in the Afrikan community. It is after a series of events- re-connecting with the masses of people of Afrikan descent in Harlem during the civil unrest, escaping from Ras the Destroyer (who cannot see that IM recognizes that the Brotherhood was manipulating both of them), escaping from the White men who are chasing him when he falls into the hole, and being in solitude for a while- that IM centers himself, finds himself, and sees his larger purpose in working on behalf and in the interests of his people. This shift is seen in the burning of the symbols of European control: the high school diploma, the letter telling him to slow down, the note with his brotherhood name on it, and the Sambo doll. It is also seen in the dream where the slave woman reminds him that freedom requires action. He is centered when he dreams that his teachers, Bledsoe, Brockway, Brother Jack, and Emerson, had castrated him, cutting him off from them, and allowing him to begin a new history for himself and with his people. Nonetheless, IM never gets to Semaj's third stage. He never re-centers himself in the survival and prosperity of persons of Afrikan descent. Rather, IM has that potential when he re-emerges from the underground.

Williams' theory (1981) contains the same Afrikan unconscious or survival thrust, and an extended self that sees the individual as part of the collectivity of Afrikan peoples. IM, as previously noted, contains some of these elements, but they are incomplete. It is in Williams' concept of "I," the expression of the collective Afrikan traits in the individual, that IM's dilemma crystallizes. IM is clearly out of rhythm with his people, and has lost his spiritual grounding as well, two aspects of Williams' concept of "I." It is the task of an identity development theory to account for these issues, and come up with solutions for them.

From Akbar's perspective, IM exhibits symptoms of alien-self, anti-self, and self-destructive disorders (Akbar, 1985b; Akbar, 2004), or, using Kambon's (2003) term, he is psychologically and culturally misoriented. Cooperating with the White townspeople at the beginning of the novel in fighting his racial brethren, being seemingly reduced to impotence while watching the naked White woman, and getting shocked chasing money on the electrified carpet, all at the battle royal, are all examples of these disorders. Letting Norton learn about the "secrets" of the campus (and of and aspect of Afrikan existence in that place) is further evidence. Being electrocuted in the factory hospital, and trained into Eurocentric beliefs on social change by the Brotherhood are further indications that he was disordered. He stays in solitude to try to think his way out of his confusion, rather than reconnecting with his family by going home or seeking support from others. It is from that vantage point that he says, in the novel's prologue, that he is invisible because others refuse to acknowledge his existence.

The I AM Model for Afrikan-centered Male Identity Development
IM has an identity, but it is mis-oriented and is in need of re-orienting. The I AM Model of Afrikan-centered Male Identity Development (the I AM Model) is based on what IM possesses, what he needs to achieve Afrikan-centered male identity development, and based on African-centered psychological theory.

Collective Afrikan unconscious. As a person of Afrikan descent, IM could possess the rhythm, spirit, bioenergy, community consciousness, and both intrapsychic and interpersonal principles which Bynum (1999) describes, but the evidence in the novel is sparse. In Azibo's (1988) and Kambon's (1998) terms, IM's inner core or ASEO is Afrikan because he is of Afrikan descent. It would be a stretch to equate who IM is with the collective Afrikan unconscious. The biogenetic nature of this principle means that it is hard to perceive. Rather it is assumed, and the outer core and behavioral component or ASC is used to clarify this issue. In the novel IM shows evidence of the communal, collective, and congenial sensibility that Diop (1991) discusses. He is part of what Fu-Kiau (1991) calls the community biogenetic rope that does not break. He is connected with his ancestors, whose memories offer him advice, and he learns from his elders, the vet, Mary, and his parents. He identifies himself as a man of Afrikan descent. Nobles and Goddard (1996) would label IM as a combination of adaptive and adoptive behaviors, each a response to the environmental impact of the world around him, and each not optimal for IM's Afrikan-centered development. The use of the dramatic consciousness technique to analogize the application of information about the nature of the collective Afrikan unconscious to a drama in which IM is a participant (Nobles & Goddard, 1993), would give IM insight into how he might think, feel, and act differently than he has in the past. Part of the drama could include incidents similar to those he encountered in the Brotherhood, for example. He could revisit the decision to leave Mary Rambo's house and move out of Harlem, the decision to move to Harlem rather than to go home when he was dismissed from the college. Or he could act out the decision to accept a different name by acting out part of a similar drama, such as playing the part of the nameless protagonist in James Weldon Johnson's (1912/1960) *The Autobiography of an Ex-Coloured Man*. The use of metaphoric memory, going from the known, a folktale or proverb, to illustrate the unknown (Nobles & Goddard, 1996), would also be useful in this context. Doing so would bring IM's actions to his conscious awareness from a different perspective, and with new force.

Kinship and family. IM was apparently connected with his family of origin before going to the college, but limited his communication with his family after he is dismissed from the college. He is dissuaded from reuniting with them by first Bledsoe and by then Brother Jack, and goes along with their wishes to his detriment. He is aided by the extended family connections that he has with the vet and with Mary Rambo, but is largely on his own. He has no friends, and has accepted isolation from his community, both when the Brotherhood demands it, and when he falls into the manhole. In the prologue and epilogue, he leaves the underground to obtain necessities, but still remains isolated. This need for family connection is a focus for work with all males of Afrikan descent who seek clarity about their identity. IM's behavior is aberrant in Nobles and Goddard's (1996) terms, is opposed to the central tenets of Afrikan cultures, and is potentially destructive for IM and his family. Here the use of mind modeling would prove helpful (Nobles & Goddard, 1993) in implementing the desired change. For example, IM could use the life of Malcolm X/El Hajj Malik El Shabazz (Clarke, 1969) to make this connection with his family. Malcolm's youth was filled with trials: His father was killed, his mother was placed in a mental hospital, and he and his siblings were separated and raised in foster homes (Howard-Pitney, 2004). Yet, the siblings remained close in adulthood. IM could re-learn the lessons of family and kinship, and observe the connections that family and community create. He could also learn the lessons that Malcolm's life teaches about who to trust and how to build an Afrikan-centered consciousness (Haley, 1976). Once IM reconnected with his family and community, the other aspects of identity development which follow will be easier for him to assimilate.

Spiritual connection. IM appears to have lost connection with the part of his Afrikan unconscious essence that is spirit. He does not pray or engage in any other spiritual undertaking to center himself when he encounters challenges. Each religious experience in the book finds IM focused on issues of more personal import to him. He is focused on his individual issues, and appears to believe that all of his life's issues are under his exclusive control. Therefore, no ancestral, spiritual, or familial connections are necessary. The deeper meanings of life do not seem to be part of how IM lives his life. Effective assistance for IM will involve spiritual reconnection. Here Nobles and Goddard's (1993) metaphoric memory would provide IM with examples from proverbs and stories about the centrality of spiritual connection. Those solutions would assist IM specifically, and Afrikan people generally. IM could then revisit past events, such as the sense he had when in the chapel at the college and a woman student sang a spiritual that moved everyone in the chapel. And he can remember how the singing of the sharecroppers, such as Jim Trueblood, moved those who heard them because of the depth of feeling and sincerity they brought to the singing (Ellison, 1947). He could also engage in culturally consistent problem-solving, by being asked to work on solutions to the issue of young people of Afrikan descent moving away from spiritual connections in favor of materialism.

Transubstantiation. At times IM operates in the best interests of his extended community, while at other times he acts individualistically, in contrast with the best interests of his community. It is hoped that he will move toward a higher level of identity and action that is indicated at the end of the novel. IM has become more self-aware, partly through his isolation and partly through the process of writing down the story of how he ended up in an underground lair that is constantly lit (in other words, writing *Invisible Man*). His stated intention is to leave the underground and assume a socially responsible role with his people. If done in an Afrikan-centered fashion, that action would initiate the growth of IM from a misoriented young male who exhibits self-destructive behavior, to I AM, a man of Afrikan descent who operates on a higher level of functioning that meets both his own needs, and those of his people. Culturally consistent problem-solving would be a good strategy to adopt in assisting IM in this area. As he came up with solutions for issues that would lead to progress for himself and for Afrikan people in the United States, he would develop a consistency of thought and action that would raise him toward the higher level of identity that he seeks. He would also benefit from image and interest discussions (Nobles & Goddard, 1993) to find the Afrikan-centered position in the discussion of personal and community issues and to respond and act appropriately. Using this concept, IM could discuss whether Afrikan-centered people would benefit from an alliance with multiracial groups and with groups that espouse socialist or communist ideologies, such as the Brotherhood. His consciousness would be raised through this process, and he would feel a deeper connection with Afrikan people in the United States.

The I AM Model is not a stage model. It can be applied, in age appropriate ways, from childhood to elderhood. It extends from the yet-to-be-born to the ancestral realm. It is centered in the beliefs, thoughts, rituals, and community of persons of Afrikan descent and within the tenets of Afrikan-centered psychologists. As each young male of Afrikan descent who seeks assistance explains his life and challenges, this model and the appropriate implementation strategies can be selected.

Implementation Strategies for the I AM Model

Kambon (1998) offers solutions for the cultural misorientation and

IM would benefit from all of them: re-education toward a Pan-Afrikan understanding and practice of history and national consciousness, and the elimination of non-Afrikan names, symbols, rituals and structures, to be replaced with Afrikan names, symbols, rituals and structures. The rites of passage models suggested by Asante (1980) and Kambon (1998) and offered by Hare and Hare (1985) and Kunjufu (1985, 1986), and naming ceremonies to mark an important event such as the journey from misorientation to Afrikan identity (Fu-Kiau) would be essential elements in this reformation. IM would also benefit from Lee and Bailey's (1998b) suggestion that he find an Afrikan-centered male role model or family of role models.

Discussion: Invisible Man and Ralph Ellison In Their Time
Ellison created a master story of possibilities. He instructs us by what is absent in the story and in IM-- the presence of parents, siblings, childhood, and his name. Perhaps IM's renaming himself is part of his resurrection, part of his rite of passage back to the Afrikan-centered world, just as members of the Nation of Islam used to drop their European-given family names in favor of an "X," symbolic of their status as a (temporarily) lost people. IM has seen the choices available to him. He descends into the underground accidentally, but while there he tries to figure out how to remake himself (Ellison, 1947). In the end, he vows to arise again and assume a socially responsible role. The young critics mistook the novel for the author, and therefore assume that IM will rise from the underground to assume a position of interpreting people of Afrikan descent, and their creative production, for people of European descent. The novel makes no such claim. IM could as easily rise to lead an Afrikan-centered effort that takes elements of the ideologies he has learned and spurned, and fashion them into a successful path to liberation. IM could easily be imagined echoing

Fanon's words:

> I am not a potentiality of something, I am wholly what I am.
> I do not have to look for the universal. No probability has any
> place inside me. My Negro consciousness does not hold itself
> out as a lack. It *is*. It is its own follower. (Fanon, 1967a, p.
> 135, [*italics in original*])

At the end of the novel, IM has yet to choose who he is to become.
That choice would be his next step.

Invisible Man is filled with folktales, and allusions to folktale
characters, survivals of the oral folktale tradition of Afrikan societies.
Brer Rabbit and Jack the Bear are mentioned, as is John Henry, a
mythic hero of the community of Afrikan descent in the United States
(Bradford, 1931). The unifying feature of these folktales is that the main
character in each of them, who symbolizes persons of Afrikan descent,
outwits and outsmarts the bigger, stronger foe, which symbolizes the
oppressive power wielded by people of European descent. Incidents
within the novel have the same folktale quality. Peter Wheatstraw is
making a living off of the used dreams of people of European descent
(the blueprints). Peter sings the blues, and talks in folktale rhythms
(Ellison, 1947). Brockway works in the lowest basement (the industrial
equivalent of working in the fields as opposed to the house during
slavery) and distrusts everyone, fearful that he will lose his job. His job,
it turns out, is to make the basis for the optic white paint. That black
dope is symbolic of the collective work and will of people of Afrikan
descent that created the economic power of the United States as they
and elements of their culture survived through group cohesiveness and
connection. In other words, Brockway is the core, the backbone of the
factory. Mary Rambo has brought the African culture of the South to

her apartment building, mothering her tenants, feeding them southern food, singing the blues, and offering them her advice and her love. Brother Tarp, escapee from a southern chain gang, another form of servitude, still has a chain link from the experience. He raises the issue of freedom by putting a picture of Frederick Douglass on the wall of the Brotherhood's Harlem office, and illustrates his own freedom by placing the broken link, a symbol of freedom attained, in IM's hand. Dupre and the other residents of a dilapidated building in Harlem burn the building down, symbolically destroying one of the master's assets, and letting the master know that those he/she has attempted to subjugate are dissatisfied and are willing and able to act to pursue their needs.

Ellison (1947) also utilizes the dozens and signifying as ways to assert the personhood of IM in the face of opposition, such as the doctors in the factory hospital, and Ras the Exhorter in the streets of Harlem. IM uses these same verbal devices to assert his authority in the Brotherhood against Wrestrum, Tobitt, and Jack. And he uses them in conversation with Peter Wheatstraw. These verbal tests of personhood are a way that people of Afrikan descent use to build emotional strength against verbal attacks, creating a steeled resolve to stand in the face of assaults launched against them. The use of folktales, the food eaten by people of Afrikan descent, the songs they made and sang (the blues, jazz, spirituals, and gospel), the sermons designed to sway a hopeful people, and Mary taking in a stranger and healing him with food and love-- all these are elements of a rich, vibrant, alive, positive culture of people of Afrikan descent. Furthermore, the novel takes place largely on Afrikan "turf": at a Black college in the rural south, and in Harlem, a Black city.

Invisibility itself is a concept consonant with the reformist school of Black psychology (Karenga, 2002). Invisibility assumes the perspective, or point of view of the other, the viewer, such as was seen in the prologue of *Invisible Man*, where IM was not "seen" by the man who bumped into him. IM eventually concludes that he is invisible to the rest of the world. But he is not invisible to all people of Afrikan descent. Some recognize him even if they do not agree with his stance on issues that are important to them. In fact, he adopts the disguises of Rinehart to *become* invisible to the Afrikan people of Harlem in order to escape the wrath of Ras and his followers. He is invisible to the white world, that external world that operates outside the Afrikan world that is the center space of the novel. By the end of the novel (in the prologue and the epilogue) IM is coming into being, or, to use the metaphor of the book, he is becoming visible-- to himself. Hence the image of living in a room flooded with light becomes significant, for the light eliminates the possibility that he will be invisible to and from himself. Symbolically and metaphorically he has seen the light. The sheer intensity of the luminescence of 1369 bulbs insists that he be seen, and that he see himself (Ellison, 1947). Hence, by the end of the novel, he is no longer invisible. He is moving from being toward being I AM, both literally and symbolically. It is this stepping back into the world, first in his own mind, and then in fact, that signals the death of externally imposed identities, and the possible re-emergence of his Afrikan identity. For himself, and for the reader, IM has become visible. By reading what he has written, and by identifying with him, the reader becomes part of IM's family by extension. As such, readers bring IM into being, into identity. What he makes of his new self is left open to the reader's speculation.

However, he is not done. He still has possibilities. When

he emerges from the factory hospital, he realizes that he is no longer afraid of White people, and that he cannot expect anything from them. During his entire sojourn with the Brotherhood, IM is subconsciously suspicious of their motives. The actions of the Brotherhood near the end of the novel prove his suspicions to be correct. His descent into the solitude of his underworld home is symbolic of an internal journey to find out who he is. As he ponders his future in his basement home, he does so with a consciousness that has been developed via the experiences he has endured, as well as through the music he hears and the dreams he has. If he chooses, he can recognize that the "crazy" veteran, his seemingly "crazy" grandfather, and Mary Rambo have offered him clues, a path to his future, guiding and mentoring him by explaining the true nature of those opposed to him, and of his place in the world. IM is the mythical and symbolic embodiment of the collective history and politics of people of Afrikan descent that are portrayed in *Invisible Man*, just as is each person of Afrikan descent in the United States, even if their ancestors did not arrive on these shores as bonds people. He is also a representation of Ausar who is murdered by his brother Set and cut into pieces. Ausar's sister/wife/twin soul/complementary polarity, Auset, re-assembles Ausar, who is resurrected after funerary rites had been properly performed, born anew (Gadalla, 2001). In the same way, IM becomes captivated by the world of White power, where he is dismembered by Bledsoe, Brockway, in the factory hospital, and in the Brotherhood. But he manages to maintain a semblance of his self, and when he falls out of the White world, the lighted world, and finds himself in a dark world, the Afrikan world, he begins to resurrects himself, and is about to re-emerge, just as Ausar did. That is, he is re-illuminated, relit.

Invisible Man also reflects the concepts of the Black Aesthetic.

It is socially relevant as well as artistically accomplished. It is socially relevant because it forces critics of Afrikan descent to address social relevance in art, and to evaluate art such as this novel from that vantage point. It also instructs those of all races that there exists an Afrikan culture in the United States that has been deadened but not destroyed despite the devastation imposed by people of European descent. And *Invisible Man* challenged other writers of the era to produce similarly socially relevant literature. *Invisible Man* comes out of the experience of Afrikan people, and is written in the blues, folktales, the dozens and signifying idioms that are central to collective Afrikan culture in the United States. And *Invisible Man* is committing. It suggests things that males of Afrikan descent must do, and things that they must not do, in order to survive, prosper, and make progress for themselves and their people. And it offers hope. IM does survive, despite the challenges that he faces. He advances and gets stronger and clearer about his identity. And, when the need arises, he fights back. In other words, he acts. In so doing, he affirms his personhood, and the collective personhood of Afrikan people. Kent (1972) is correct: Karenga's (1980) three-part framework-- social relevance, collectivity, and commitment-- is limiting, and purposely so. It places works by Afrikans in the United States within the context of liberating themselves from continuing oppression by offering an Afrikan paradigm for creativity. In so doing, it limits Eurocentric theories and paradigms from consideration, and creates a paradigm for the development of an Afrikan-centered identity.

Invisible Man and Identity Theory: Review
Of what does this Afrikan identity consist? Being grounded in the idioms of people of Afrikan descent is one component. It involves moving to the rhythms of the music of people of Afrikan descent, as IM is swayed by the blues and spirituals. It consists of being part of

an extended Afrikan family (Smallwood & Shields, 1997), as Mary exhibits when she takes IM into her "family." It involves the types of food eaten, and the communal and spiritual aspects of consuming that food together. It involves living up to a family and personal name, as illustrated when Mary Rambo names IM after Afrikan heroes to inspire him. Naming, in this sense, confers identity (Ssensalo, 1978). All of this equates to a grounding in Afrikan American culture, and reflects Afrikan ritual (Some, 1993). Afrikan culture requires a commitment to recognizing the challenges that people of Afrikan descent face in America, and addressing oneself to those challenges, to the extent of one's abilities and talents. In other words, it involves translating conscious thought into conscious action, and conscious action into conscious thought.

IM does not appear to be conscious of and grounded in his Afrikanity, his connection with and immersion in Afrikan culture during his high school days. He knows that he is Black and that Whites controlled the economics and wealth of his home town. He has the same perspective when he is at the college. His behavior in these phases of his life can be described as a combination of adapted and adopted behavior, using the Nobles/Goddard framework or a combination of anti-self and self-destructive behavior, using Akbar's framework. IM develops an awareness of his Afrikanity through his experiences in New York, where he meets people of Afrikan descent from other parts of the world and from other cultures, such as Ras, who is from the West Indies. He also learns about Marcus Garvey and the Universal Negro Improvement Association, with its "Race First" and "Back to Africa" orientations (Clarke & Garvey, 1974; Martin, 1976), but does not identify with those orientations. IM appears to understand that he is an Afrikan man living in the United States during his conversations with Tod Clifton, though he rejects the position advocated by Ras the Exhorter on the issue. IM adopts the credo of the Brotherhood, and repeatedly is pushed into adaptive behavior as he fluctuates in and out of the Afrikan culture of Harlem while working in the Brotherhood. His Afrikan-ness is growing in remembering and longing for life as he lived it in his younger days, his connection with the people of Afrikan descent in Harlem, and in recognizing the alien qualities of the members of the Brotherhood.

Culturally, IM is Afrikan throughout the novel, though he often does not recognize it, or act in accordance with its dictates. In other words, he has yet to become Afrikan-centered. His understanding of the students singing in the gospel choir at the college is deeper than the surfaces of the music: he reacts authentically at the plaintive tones and timbres of the singing. He goes back to his roots as he eats a yam on a cold city street, once again grounded in the culture of his youth. He uses folktales and folklore, both verbally and in his imagination to avoid further damage in the factory hospital. This he does instinctively, an Afrikan unconscious behavioral response to being attacked (electrically shocked) while defenseless, strapped to a table. He respects Brother Tarp and the chain gang link that Tarp gives him, and sees the connection between it and the picture of Frederick Douglass, the one-time slave, that Tarp hangs on the wall. His anger at the Sambo doll bank in Mary's home, and at the Sambo doll that Tod Clifton is selling, is an Afrikan American response. Both Sambos are symbols of an accommodation to oppression that results in a type of survival that he rejects, though, for a time, he is controlled by the Brotherhood in seemingly similar fashion. When he speaks at the eviction of the elderly Afrikan American couple in Harlem, for the Brotherhood during his first speech in Harlem, and at Tod Clifton's funeral, he speaks in the call and response style that is evocative of Afrikan American communal oral traditions and group communication (Neal, 1968). His clarity in the prologue, amidst a drug-induced dream, about the spiritual he hears, the slavery stories he dreams, and the Louis Armstrong blues that frames his experience, is consistent and reflects genuine real strivings and emotions. His cultural acuity would need to be refined, and perhaps refocused-- made more authentic, and centered in Afrika.

Another issue that IM faces is a lack of relatedness. After studying the Kemite origins of psychology, Akbar (1994) concluded that: "the human being is transpersonal and essentially connected with the divine and everything else in nature" (p. 63). This idea is consonant with Noble's concept of an extended self (Nobles, 1972). IM is alone through most of the novel, an example of aberrant behavior which, if not corrected, would destroy him eventually. He has chosen, wisely or not, to "go it alone" and not involve his family or friends in the trials that he encounters. He makes this decision twice through the course of the novel, and in each instance it has significant consequences for him. The first time occurs when IM threatens to tell his family that Bledsoe is kicking him out of the college. Bledsoe was fearful of the reprisals that could come if word got out about the injuries suffered by a trustee while on the campus, about the existence of the Trueblood family and their "story," and about the brothel on the campus. Bledsoe is more concerned about IM telling his parents about what has occurred than he is about IM telling Norton (Ellison, 1947). Bledsoe's fear is based on his understanding of Afrikan culture and European culture, and that he has transgressed against both in his running of the college. To distract IM from informing his parents, Bledsoe offers him letters of introduction to find a job in New York, thus steering IM away from any desire to reunite with his family. IM trusts Bledsoe, overriding his instincts to reconnect with his family, a clear cultural and practical error. The second instance occurred when IM joined the Brotherhood. He was given a new name by which he was to be known, and was told to move into a new apartment. During the process, Brother Jack suggested that he cut off all contact with his family. In agreeing to cut off communication with his family, IM acts against his own best interests and in line with the interests of representatives of an alien culture, further isolating himself. At that point, he was without family, friends, and extended family, since he had moved out of the apartment of his surrogate mother, Mary Rambo, in Harlem. His fleeting connection with the Afrikan community in Harlem, gained when he defended the elderly couple against eviction, is severed. When he ends up in the isolated underground space, under Harlem but not in it, he has forsaken everyone.

Lee and Bailey (1998b) see a lack of role models for males of Afrikan descent as one of the causes for insufficient male identity development. Their solutions to this absence include providing such male role models, incorporating the legacy of strength, pride, love, kinship, respect and stability that have characterized Black families, and incorporating African and African American culture, including music and poetry. IM would need to be reconnected to family, his racial family and his biological family, to be reconnected to and with his friends, and to be reconnected to and with his community. For only by so doing would his individual self be reconnected with his extended self and his kinship network, refueling his personhood in the community context that is an essence of healthy persons of Afrikan descent.

IM is partially connected with his ancestors. He repeatedly reviews ancestral wisdom from his grandfather, each time seeking further insight and truth, but IM never quite understands what his grandfather means. And he dreams of a woman involved in the holocaust of enslavement. She tells IM that she is there to rid him of his ambivalence when he asks her what freedom was. She could be interpreted as an ancestral spirit come to give him information to assist in his survival and growth (Ellison, 1947). Again, IM does not quite understand her. IM is not spiritually connected. He also shows no signs of being religiously connected. He requires spiritual instruction and guidance to reconnect him to/with a form of spirituality consistent with Afrikan culture. His attempts to understand his grandfather's messages are authentic behaviors, part of his search for his authentic self.

IM is connected to the music, the spirituals, gospel, and the blues, but is often out of sync with the rhythms of Afrikan American life. Bynum (1999) asserts that rhythm is an organizing principle that allows individuals to integrate with others, resulting in a shared identity. IM struggled to understand Peter Wheatstraw when Peter spoke to IM in signifying terms. When he entered a bar near the end of the novel, he cannot seem to connect with the people there, people from his own community. IM also rejected the opportunity to enter Rinehart's church when he found it. His behavior in these areas is maladaptive and self-destructive. He has adopted a Eurocentric framework to exist within the Brotherhood, and that is why he is slightly out of sync with his cultural context. However, during the civil disturbance that culminates with him falling into the manhole, he merges with the masses of people of Afrikan descent, going along with them as they avoid the police and burn down the dilapidated apartment building. He becomes connected with his community, and is temporarily immersed in their lives, swept along with them, rather than attempting to lead them somewhere. Here he resolves the problem of adaptive, anti-self disordered behavior, and expresses authentic Afrikan behavior: righteous indignation and response to oppression. This result points to his need to continue authentic behavior, but in an Afrikan-centered context, in order to be fully reintegrated into Afrikan life once he leaves the underground.

On other cultural levels, IM has lost himself as well. He is steeped in Afrikan American language idioms: he speaks and understands an Afrikan form of English, but does not use it often, and is initially unable to understand Peter Wheatstraw when he uses it. Through the course of the novel he does not engage in any of the rituals of the Afrikan community. He does not sing, he does not dance, he does not seek spiritual connection, and is absent from Harlem when cultural celebrations occur. In these ways, he was isolated from his own community, and from himself as well, behavior that is aberrant, or self-destructive. His physical isolation in the apartment owned by the Brotherhood was symbolic of his separation from his family, from his community, and from his culture. He would need assistance in reconnecting with and reclaiming these connections.

Finally, IM has few intimate relationships in his life. In this way he is also disconnected. He lacks family, friends, associates, and love. He has no one to trust, no one to be a part of, no group in which to center himself. In this respect, he appears to be out of balance, and out of control. He would benefit from assistance on this issue aimed at reconnecting him with his esteem, worth, and emotional center.

The I AM Model offers insights about what the components of an Afrikan-centered identity are. It offers a way to understand how to take a young Afrikan man who is misoriented, and to conceptualize his problems. Nobles and Goddard (1993) offer a set of culturally connected pathways by which such a young man can be reconnected with those aspects of his Afrikan self that have been confused. The process of reclaiming an Afrikan identity is not solely the province of a single person, such as a therapist. The involvement of family in the healing and in rites of passage is vital. The assistance of members of the young man's age cohort who are themselves centered in their Afrikan identity is also of prime importance. The goal is that each such young man emerges confident, connected with his family, his people, with his past and his future. And the connection to other young men and women who are similarly Afrikan-centered will create a community of people working toward the same goals. When this occurs, the trials of daily living will occur in a perspective that will allow him to live with confidence that these are but momentary bumps on the road to the larger goal of the freedom and advancement of Afrikan people in the United States and around the world.

IM ends the novel by saying: "And it is this which frightens me: Who knows by that, on the lower frequencies, I speak for you?" (Ellison, 1947, p. 581). The process of identity development assisted by the I AM Model seems to answer the question as to whether the novel speaks to people of Afrikan descent.

CHAPTER V
Conclusions and Recommendations for Further Research

We must start by acknowledging the fact that we might have to step back in order to step forward. We might have to go back to what I call the fork in the road where we misread the sign boards, and once we read the sign boards correctly, and find the arrow pointing toward unity, self-reliance, and Pan-Africanism, this might be the road that leads us home. Travelling down that road, we must restore our humanity first; we will be changing the world by changing ourselves first. This might be our holy mission. It might be the legacy that we can leave for our children, and their children still unborn. (Clarke, 1999, p. 98)

Whatever sickness we have becomes our teacher. Now your job as a healer is to help people to want to learn from their own life teachers . . . not to take those teachers away so they don't have to do their own work. No matter what you try to do, their work is their work. (Alexander, 2005, pp. 88-89)

Conclusions

Literature uses myth and symbol to illustrate life. Myths and symbols are also used in other areas of life and culture, such as religion, politics, and psychology. Afrikans use myths, symbols and literary forms to enhance, explain, and clarify their collective existence. Afrikan cultures from the ancient Kemites (Carruthers, 1995) to the Dogon (Grills, 2004a) speak to the power that spoken words and speech have.

Rowe (1995) adds a modern interpretation of the use of language as a symbolism to be used to communicate concepts. Extending these concepts from the spoken word to the written word, one can see how spoken symbols can be transferred to written form to communicate powerful and deep knowledge. This dissertation combines the symbols of literature with the symbols of Afrikan psychology to create a deeper understanding of identity development. Milliones modified the Cross and Thomas models of Afrikan identity development by culling information from non-fiction literature written by and about persons of Afrikan descent (Marks, Settles, Cooke, Morgan & Sellers, 2004). Dixon (1976) believed that proverbs, folk tales, and other Afrikan cultural expressions reflect the collective experience or judgment of the Afrikan community as a whole. Harper-Browne (1996) proposed that Afrikan American psychology is revealed in Afrikan American literature should be part of a curriculum in Black psychology.

The culling of Afrikan psychological theory from Afrikan creative work is of ground-breaking import. The last generation of Afrikan-centered scholars created an Afrikan-centered psychology, and expanded the critical framing and analysis of Afrikan American literature. This movement is operating parallel to that of Afrikan scholars on the Afrikan continent who are rejecting the European frameworks of philosophy and are arguing for Afrikan-centered philosophical concepts and praxis (Serequeberhan, 1991). The current generation has expanded both arenas, creating Afrikan-centered organizations in the United States both psychology and literature. Both Afrikan American psychology and Afrikan American literature are taught on college campuses. However, scholars in any Afrikan-centered endeavor are challenged by the inability to find publishers to publish their works. And when mainstream publishers do publish such works, the works are not publicized, making access to their current research more tedious and difficult. However, Afrikan-centered presses have been established and have addressed that issue, and some of the hallmark documents in both disciplines have been published by those presses. Now a new generation has arrived, and new challenges face Afrikan Americans in general, and Afrikan American psychologists in particular.

The task of Afrikan reclamation of their historic greatness that started with Delany (1879/1991), Blyden (1908/1994, 1888/1994), and DuBois (1978, 1946/1965), and which continued with Diop (1974, 1987), ben Jochannan (1971, 1981), Clarke (1969, 1991, 1993, 1994, 1999), and others, continues in the present day with Browder (1992), Carruthers (1999), Suzar (1999), and Jubal (1991). In *Afrikan Alchemy*, Dismukes (1995), for example, uses *The Kybalion* to frame a destination of love and healing for Afrikan people. Authors of European descent, such as Moore (2001), Jensen (2005), and Wise (2005) are also making the journey into what Jubal calls the "Black Truth." And Afrikan scholars are combining two or more distinct European academic disciplines to produce larger truths, just as this piece unites Afrikan literature with Afrikan/Black psychology to expand Afrikan-centered psychological theory. For example, in a series of law review articles, Herbert (2002, 2003, 2005) has used Afrikan culture, history, and literature with legal doctrine to illustrate how rights granted to people of Afrikan descent under the Constitution have been curtailed by interference from law enforcement. She demonstrates, for example, how bad treatment by police of parents of Afrikan descent in the presence of their children, replays a situation found in the holocaust of enslavement, where such parents were not allowed to parent their children. Their parental responsibilities were removed from them. She points out that in such circumstances, the bonds of kinship are cut in favor of the bonds of ownership (in the enslavement experience) and control (in the case of modern policing). In so doing, she adopts Dyson's (2004) term "un-kinning," the dismantling of the relations within a family that binds it together, to describe the results of these actions (Herbert, 2005). Ephirim-Donkor (1997) explains how the Akan people of West Africa are a product of their mother's lineage, and without that lineage, the individual is cut off from her or his community. Herbert's unkinning echoes this process. This pairing of Afrikan cultural processes and law holds promise for further clarifying the places occupied by persons of Afrikan descent, and the spaces left by their forced sojourn in an alien cultural context.

As Christian (1969) rightly notes, " . . . actually the hero has found his identity before the book opens. His first statement is to assure us that he knows who he is. I am an invisible man,' he proclaims . . . " (p. 357). He is clear that he is invisible, to people of European descent. In fact, he is too visible to some of his own people, who equate him with the Brotherhood that has betrayed them. Writing his story was his initiation back into himself. IM's challenge now is how to be in the context of his people. He lacks family, village, and nation. Having sought and found who he is and is not, he now needs to know who he is and is not in the context of a family, village, and nation. In this respect, his life is a metaphor for Afrikan people. Obenga has taken the traditional Afrikan question "Who am I?" and responded by asking "Who are we?" (Carr, 1997, p. 287). In so doing, he contextualizes an individual such as IM with his people, his family and his community. Afrikan people in the United States also need to re-find who they are first, and then find themselves in the context of America. It is this search for who IM is in terms of his people that produces a framework for the model of Afrikan-centered identity development.

From the scholars of ancient Kemet to today's Afrikan-centered scholars, it is believed that knowledge of self is the key to all knowledge (Myers, 1991). Reading about IM's identity struggles allows the reader to reflect and analyze their own identity struggles, if they wish. It is commonly held that creative arts is a springboard for change in all societies. Afrikan America is no different. Keys to understanding the varied characteristics of Afrikan American people can be found in literature. Those characteristics, or types, can then be analyzed, and solutions created to provide a psychological health that places the person in an Afrikan American context. Afrikans are holistic. That is, Afrikans perceive life in its entirety in order to make sense of it. They do not compartmentalize it to find its essence. Using Afrikan literature to develop Afrikan psychological theory is also holistic, taking two disciplines and combining them to make a more inclusive discipline.

The goal of Afrikan-centered psychology is not to rehabilitate Afrikan Americans to resume a subordinate role in this society. Rather, this psychology has as its purpose and goal to heal persons into a conscious self-awareness that will render them strong in the face of oppression, ready and able to define, defend, and drive themselves toward a fuller life unencumbered by the perspectives, power, and persuasions of White supremacy. Baruti writes: "We do this by taking the time to deeply and courageously think just what institutions or actions Afrikans *need* to become truly liberated and making that into a reality (2005, p. 159, [*italics in original*]).

Boykin, Franklin and Yates (1979) extend the definition of the identity of persons of Afrikan descent to psychologists of Afrikan descent. They see identifying oneself as a person of Afrikan descent as central, rather than incidental, to being an effective psychologist for persons of Afrikan descent. This is in line with their professional goal of improving the lives of persons of Afrikan descent. Phillips (1996) believes that it is through the therapist that healing occurs, rather than healing being caused by the therapist. This idea is in concert with the groupness and collective consciousness that is central to the Afrikan worldview. Dennard (1998) suggests that therapists working with persons of Afrikan descent must search for the strengths in the person that resonate with Afrikan life, rather than focusing on pathology. In so doing, Dennard sees the therapist using Azibo's (1988) nosology as a framework, to take the client's awareness of his situation and reframing it, contextualizing it as part of a condition common to members of his racial group. In so doing, the individual is made part of a group, and the client finds renewed energy in the sense that he is not alone. Equally as vital is the sense that the therapist identifies with and understands the client in the context of Afrikan culture operating in the setting of the United States.

Afrikans in America, like our European counterparts, have separated ourselves from nature. We live in enclosed shelters, in cars, have artificial floors, and wear shoes to avoid touching the earth with our feet. We avoid the sun by the use of shelters with lights, and cover ourselves from exposure to it with clothes. We avoid the wind and the air by enclosing our shelters and using heat and air conditioning. We avoid natural water. We have chemicals in our showers, baths, and pools to "clean" our water. While so doing, we pollute the air, block the sun, pollute the water and pollute the earth. We have much work to do to reunite ourselves with these, our physical roots, as we engage in the process of reconnecting with our cultural and spiritual roots. *Invisible Man* is both an illustration of and a metaphor for a "healing conversation." The story of IM's life is a way of gaining the deep connections that are the goals of healing (Rowe & Webb-Msemaji (2004).

Recommendations

The I AM Model shows that literature by and about people of Afrikan descent can be utilized to provide a path to Afrikan-centered identity development.[23] From this beginning, intervention strategies that have come out of the Afrikan-centered psychology movement must be applied to this model and analyzed to see if they can be of use in healing persons of Afrikan descent. As with any theoretic model, studies must be conducted using the model to assess its viability, and to alter it in light of new evidence. If the I AM Model has avoided the errors of which Azibo and Robinson (2004) speak, then application of the model in therapeutic settings with males of Afrikan descent will prove beneficial.

It appears that the model can be used on two differing levels. One is as a metaphor for a therapeutic situation, where, as in the mind modeling and metaphoric memory techniques proposed by Nobles and Goddard (1993), it could be used to capture or recapture lived experience to determine if and where the patient is misoriented. For example, if one is working with a young man who is disconnected from his family, the *Invisible Man* story could be used to assist him. The details of IM's choices not to unite with his family after his problems at the college and when he joined the Brotherhood would be a way to discuss these family and kinship issues. The other level would be to use the model to deduce issues to be addressed in a therapy situation. This method could prove useful in preschools and elementary schools, where children of Afrikan descent are often miscast, misunderstood, misdiagnosed, and misdirected. Utilizing rituals and rites of passage for the movement of male children of Afrikan descent from home to preschool, from preschool to elementary, from elementary to secondary school, and so on, would provide these young people with goals, direction, and modeling by older youth who complete the process. Adult and elder males can go through these rites of passage and rituals, to re-establish an Afrikan-centered community among all age groups. This process would also be useful in reclaiming those males of Afrikan descent who have been incarcerated in the United States. By building age cohorts, groups of males of Afrikan descent of the same age group who go through the rites together, a community of men is grown whose obligation is to look out for each other and for the Afrikan-centered values that the I AM Model offers as guideposts.

The story of I AM in the novel points to the centrality of providing males with new names as signifiers when they transition to a new stage of development (Holloway, 2006). Many of the Afrikan-centered theorists cited herein have chosen Afrikan names to add to their given names, or to add to their Afrikan names. Some have adopted Afrikan first, middle and last names to signify their new centeredness in Afrikan realities. Naming, whether after famous persons of Afrikan descent, or after attributes possessed by the person that are admired, should be a feature of these rites of passage. Further study should be done concerning the effect on the identity development of an Afrikan person who chooses an Afrikan name, or has an Afrikan name chosen by elders in the group.

Franklin (1992) notes that males of Afrikan descent are reticent to attend therapy, fearing the vulnerability of personal disclosure. Franklin (1998) uses the term "Invisibility" to signify the rejection of the identity of males of Afrikan descent by mainstream society and the consequent covering up of their true self by males of Afrikan descent in response. For these men, positive validation is absent from most interactions with the larger society. Franklin posits that racial identity development is an antidote to this invisibility. Boyd-Franklin and Franklin (2000) suggest a number of ways that young males of Afrikan descent can establish a positive identity: have them learn about Afrikan American history; praise their accomplishments; provide them with good role models and mentors to counter any negative peer pressure; institute rites of passage programs for young males transitioning to manhood; involve the entire family in the process; focus on spirituality; and discuss issues around sex and sexuality, drugs and gangs. In her Multisystems Model, Boyd-Franklin (1998) combines individual family members with extended, spiritual and community family resources in discussing the causes and cures for issues faced by families of Afrikan descent. Franklin (1998) advises therapists to allow those males of Afrikan descent seeking therapy to express their distrust of the process. He notes that if those males have been accustomed to distrusting others and protecting themselves against vulnerability, then they must be given time to become comfortable with the therapeutic situation before the therapy can progress. The I AM Model could assist in this process, adding depth to family involvement by addressing kinship in the broader context. This discussion might provide the young man with an understanding of the ancestors from which he springs, thereby making a "place" for him within the larger group. He may also better understand the gifts and talents he has inherited from his ancestors, and know his responsibilities to his family and community. The I AM Model also could provide the young man with clarity about his need to act in the best interests of his community as a way of acting in his own self-interest, which Nobles and Goddard (1993) called transubstantiation.

Jenkins (1995) utilizes a humanistic framework to discuss issues relating to Afrikan people in America. Jenkins focuses on raising the level of an Afrikan male's racial self-concept, assisting him to assert his humanity without striking out aggressively against those persons and institutions that oppress him. Jenkins notes that the taking on of a new and positive racial group identity by persons of Afrikan descent assisted in raising the self-esteem of males. This process requires an understanding of how the individual understands and conceptualizes what is transpiring in them and around them, and how they act based on the choices that they see as available to them (Jenkins, 2004). While he acknowledges a place for a group-centered orientation, Jenkins focuses primarily on individual identity or "personal agency" (Jenkins, 1998, p. 182). The focus of therapy using this framework is to move the young man from "distorted, self-defeating, and unproductive" behavior (Jenkins, 1995, p. 247) toward productive behavior. He notes that religiosity is a key ingredient in this process of individual reclamation. The I AM Model would provide a valuable perspective in this framework as well. Broadening Jenkins' concept of the individual to include his place as part of a family and community, and as part of an extended self that reaches back to ancestors and to those yet to be born as part of understanding each young man seen in therapy, both the therapist and the young man would develop a clearer understanding of who the young man is and how he fits within the larger framework of his family and community. The sense of a shared connection that the collective Afrikan unconscious implies is another avenue of inquiry and discussion designed to assist the young man on this journey. Broadening the concept of religion into the larger framework of spirituality would also fit the humanistic frame, as it universalizes the feelings, thoughts, aspirations, and understanding of the connection between the individual and the rest of life. And the addition of an understanding of the possibility of moving to a higher level of personal and group identity (transubstantiation) could well inspire a young man to work harder toward goals that benefit not only him but also members of his family and community. This result might replicate the importance attached to ideology and spirituality found by Edwards (1999) in a study of a cross-section of Afrikan American men and women.

Watts, et al. (1999) created a sociopolitical model for understanding male identity development. This theory views the drive to achieve racial self-esteem and spiritual commitment to a liberating set of beliefs as keys to structuring a new racial identity that does away with the victimization of oppression. This is a stage model where the final stage, the liberation stage, involves taking action to eliminate oppression on both the individual and group level. To do so, one must build what is called "critical consciousness" (Watts, et al., 1999, p. 264), a method of critical thinking that involves coming up with constructive solutions for problems. The I AM Model could expand the sociopolitical nature of this model by providing the kinship connection that expands problem definition and solution to the family, community, and racial levels. The I AM Model also places such a sociopolitical model in context, centering it on the collective Afrikan unconscious, a concept that clarifies the collective nature of liberation and struggle against oppression by people of Afrikan descent throughout the world and across time. The model expands and clarifies the nature and definition of spirituality. And it discusses the group-centered nature of Afrikan male identity enhancement (transubstantiation) by moving the focus of change from the individual to the individual acting as part of a group, community, and/or nation.

A study of models of Afrikan American racial identity by Marks, et al. (2004) notes that most such models are drawn from the perspective that people of Afrikan descent develop their self-concept based on how persons from other races view them, rather than from their own family, friends, race, and those in their environment. The study discusses the Sanders-Thompson Multidimensional Model of Racial Identity, which has physical, cultural, sociopolitical, and psychological components (see Sanders-Thompson, 1995). The I AM Model could expand the effectiveness of this model by adding family and kinship, spirituality, and transubstantiation components to the model. Such an outcome would resonate with the finding in a study by Chavous, Rivas, Green and Helaire (2002) that strong racial identification is correlated with success in academic performance at a predominantly white university.

Myers (1993) offers an analysis of the Afrikan-centered worldview in her work on Optimal Psychology. This work perceives both cultural differences and human similarities to be important for identity theory development (Myers & Haggins, 1998). Optimal theory is defined and framed on the seven principles of *The Kybalion* (see Myers, 1999, pp. 319-321). Consciousness, the person's conceptual system, is central to optimal theory. An optimal conceptual system unites the spiritual and the material into one coherent worldview. A male of Afrikan descent who is socialized into and accepts a less than optimal system of concepts, a suboptimal system, struggles to find meaning in himself and in his world. The Eurocentric worldview is a suboptimal system for persons of Afrikan descent. An optimal or holistic identity is achieved when the person retakes or regains his own self-worth. Another way of defining this is that males of Afrikan descent frame their own character on their own terms. Myers framed a six-phase model of identity development utilizing optimal theory. The phases are collectively conceived as an expanding spiral rather than as a linear construct. The phases are: an absence of conscious awareness; an individual awareness framed by those who raise the person; dissonance when others question aspects of the person's self, leading them to question themselves; immersion, connecting with others who are similarly misperceived; internalization of feelings of positive self-worth; integration, using the new self-concept to re-evaluate their understanding of themselves, others, and the world at large; and transformation, the defining of oneself personally, familially, and communitarily, and as one with ancestors, the future generations, and nature (Myers & Haggins, 1998). According to Myers, this theory is applicable to everyone regardless of race. The theory is utilized with a therapeutic strategy called "Belief System Analysis" (Myers, 1999, p. 331) whose goal is the transformation of consciousness. This strategy is similar in concept and goal to that of transubstantiation in the I AM Model. The difference is in the I AM Model's focus on males of Afrikan descent, whereas Optimal theory is applicable to all, regardless of race or gender. The addition of the concept of a higher or more clear Afrikan male identity, as seen in the I AM Model, would enhance the effectiveness of this transformation of consciousness when the patient is a male of Afrikan descent. The optimal theory model addresses spirituality and family/kinship concerns, and is consonant with the I AM Model in those respects. And in its use of the principles of *The Kybalion*, optimal theory evidences a belief in Bynum's (1999) collective Afrikan unconsciousness.[24]

Philips (1998) has created an African-centered therapeutic system, called NTU psychology, that utilizes terms derived from Bantu philosophy and which resonate with and may be derived from Tehutian philosophy as found in *The Kybalion*. This system views all aspects of life as energy which is expressed as matter, emotion, or spirit. The NTU principles, which are used for diagnosis, consist of establishing harmony within the patient, and between the patient and their world, balancing seemingly opposing forces in life, recognizing the energetic interconnection between human beings, and finding the patient's ability to express their truth (Gregory & Harper, 2001). This theory posits a sixth sense, called the bodymind, that is akin to the concept of intuition in the West, that operates in concert with the other five senses to provide the individual with an understanding of the world and the truths to be learned from and in it (Philips, 1998). The phases of therapy include harmony, awareness, alignment, actualization (the activation of potential), and synthesis, leading to internal balance and more fruitful behavior. Once the patient has achieved synthesis, they are invited to live within the principles of Maat (Beatty, 1997) and within those principles found in the Nguzo Saba (Karenga, 1967). Kwate (2003) tested the effectiveness of the Nguzo Saba as used in the Africentrism Scale created by Grills and Longshore (1996). The results validate the use of the Nguzo Saba in therapeutic settings with persons of Afrikan descent, and show identity to be an ongoing process that evolves with age and knowledge. The I AM Model fits well within the Afrikan-centered framework of NTU psychology. The collective Afrikan unconscious (Bynum, 1999) is an energy, a spiritual center that inhabits all persons of Afrikan descent, and is in line with the Bantu basis for NTU. Utilizing the I AM Model's focus on kinship and family will be of service in establishing harmony, awareness and alignment on both the personal and interpersonal level for those males participating in NTU therapy. The result could be a synthesis that involves the individual in relation to self, family and community. More importantly, NTU offers the clearest connection with the ancestral realm and to those yet to be born, as it views all as energy and spirit, including the continuity between past, present and future life, of which the patient is part. Fu-Kiau (1991), himself of Bantu heritage, calls this continuity the biogenetic rope that does not break. The spiritual aspect of the I AM Model is intrinsic to NTU psychology. The final result of this technique, synthesis, requires living at a higher level of identity. The usefulness of the concept of transubstantiation is that it would fit the patient into a place where this higher level of humanity is achieved in a group and community context, as well as individually.

Azibo (1988) created a classification system to describe the Afrikan personality. Noting that personality order is a prerequisite to understanding personality disorder, Azibo used an Afrikan-centered worldview to construct his system. The three disorders see in persons of Afrikan descent in the United States, and derived from the biogenetic core of Afrikan people, are psychological misorientation, where the Afrikan genetic core is disrupted by operating without Afrikan-centered beliefs; mentacide, accepting a Eurocentric worldview and believing that they are psychologically unbalanced; and other Afrikan personality disorders that spring up as a result of psychological misorientation. Dennard (1998) used Azibo's categorization system in therapy with a man of Afrikan descent. Dennard reframed the man's understanding of his Afrikanity by grounding him in the conditions common to people of Afrikan descent in the United States. The addition of the concept of the collective Afrikan unconscious from the I AM Model could have added additional clarity to that discussion. Dennard reduced the patient's "WEUSI anxiety" or Eurocentric individualism (Dennard, 1998, p. 188), by pointing out how the patient's business partner and his secretary became his friends and part of his healing. This is consonant with the I AM Model's concept of re-establishing an Afrikan family and kinship framework to center males of Afrikan descent and create a place for them within a community context. In a study by Tyler, Boykin, Boelter, and Dillihunt (2005), Afrikan communalism, the sharing with others of Afrikan descent, was found to be a strong ethos in persons of Afrikan descent. Using the I AM Model to enhance this man's spiritual understanding and connection could have resulted in his attaining additional self-confidence, and soon he might recognize himself as acting with a clearer, more well-defined sense of who he was (transubstantiation). In summary, the I AM Model enhances current models for assisting males of Afrikan descent in re-centering themselves by broadening the scope of analysis and providing additional levels of Afrikan cultural experience to ground them on their journey toward empowerment and identity clarity.

The field of identity development has, to date, been overly wedded to one set of theories, those of Cross (1979), Thomas (1971), and others. The I AM Model stands as a critique of that theory, and can be used to re-evaluate identity development literature from an Afrikan-centered frame. Future work on Afrikan male identity development would benefit from the view of Holdstock (2002), who believes that psychology in sub-Saharan Afrika can be best understood through forms of aesthetic expression such as art, dance, poetry and theater produced by Afrikan persons in those locales. Study of communities of Afrikan descent in South and Central America, in India, Australia and New Zealand, and in the islands of the Pacific could yield further insight into the nature and process of Afrikan-centered identity, its problems, processes and solutions.

One obvious limitation of this work is that Ellison's work does not lend itself to an analysis of Afrikan American women. However, if this model proves successful, it can be extended to other groups of people of Afrikan descent by means of folklore, music, religious observances, other creative arts, and other elements of Afrikan cultural expression.

Members of the Association of Black Psychologists have been engaged in an ongoing discussion over the efficacy, validity, scope and direction of Black and Afrikan psychology for the past 2 decades. This discussion is ongoing (Fairchild, 2004; Grills, 2004a; Grills, 2004b). There also exists an ongoing discussion between the generations of scholars who populate the community of the Association of Black Psychologists over perspectives on what constitutes Black and Afrikan-centered psychology, and who defines and elaborates it in theory, operation, and practice (Ajamu, 2004; Grills, 2004a; Grills, 2004b). Obasi, Prince, Bolden, Richardson, and Walker (2005) continued an oral discussion that had begun at the 2005 annual conference of the Association of Black Psychologists on its members' ideological, ethical, practical, and future leadership orientations, by encapsulating the position of younger Afrikan psychologists, followed by responses from others who had been involved in those discussions. These discussions have consequences for the direction in which identity development is defined, perceived, and practiced by Afrikan-centered psychologists. As the scholars of the three Egbe (Ajamu, 2004), or generations of Afrikan psychologists grapple with theory and practice, further research on Afrikan identity development will, of need, have to encompass all that culturally emanates from people of Afrikan descent, including novels such as *Invisible Man*. One hopes that these scholars heed Kelley's (2004) call for intellectuals to create real change for people of Afrikan descent. *Invisible Man* is useful in this process and not confined to its own epoch because the issues IM confronts are faced by all Afrikan males in the United States who are on the path toward Afrikan-centeredness.

One of the arenas that requires further attention and which is beyond the scope of the present effort is the development of a range of modalities to be used to treat the ills of persons of Afrikan descent in this society. To date, the use of mainstream psychology, maintained as it is on a concept of science that is consonant with a European worldview, has had little positive effect on persons of Afrikan descent. As Gordon (1995) notes, racism is concealed in a society such as the United States by embedding its values so deeply in the structure of the society that racist values become familiar, and therefore "invisible" (p. 38). Wedded as this mainstream psychology is to a system of delivery and education that results in few Afrikan Americans being trained as psychologists, and few Afrikan Americans receiving psychological services, alternatives must be created.

The I AM Model would be elaborated further if male characters of Afrikan descent from other novels were included. For example, the character of John Grimes and his quest for identity and definition of his sexuality in Baldwin's (1948) *Go Tell It on the Mountain* could be used. Richard Wright's (1937) quest for selfhood in his autobiography *Black Boy* could be added. And James Weldon Johnson's (1912/1960) novel about an anonymous character who is of Afrikan descent but decides to pass for white in *The Autobiography of an Ex-Colored Man* could be used to discuss the tensions involved in adapting to versus adopting an alien culture.

More difficult issues also remain to be resolved. Chief of

these is the place that the concept of race occupies in Afrikan identity development. Prior to the incursion of the West on people of Afrikan descent, race was an irrelevant issue (Chinweizu, 1975). Europeans created and enforced a concept of race that is extant in most of the world occupied by persons of Afrikan descent (Fredrickson, 1988; Smedley, 1999). Massive miscegenation of peoples of European and Afrikan descent, both forced and consensual, has occurred over the centuries. As a result, the physical distinctions between races has become blurred (Sollors, 1997, 2000; Williamson, 1984). Afrikan-centered scholars have long grappled with this externally-imposed quandary raised directly by Davis (1991) in the title of his book: *Who Is Black? One Nation's Definition.* Kambon states that the generally held view is that as long as the person of Afrikan descent has identifiable Afrikan physical characteristics, then they should be accepted as being Afrikan. If they do not have these identifiable characteristics, described by some as being able to "pass" for white, then they should not be viewed as being of Afrikan descent (Kambon, 1998). Nobles (1998) addressed this issue in a broader way than did Kambon. He noted that being Afrikan-centered is a cultural expression of a *"quality of thought and practice"* (p. 190, [*italics in original*]) that is rooted in people of Afrikan ancestry. This view expands the narrower issue of race into the broader view of culture. In so doing, the issue of race recedes and the issue of culture assumes its place as central to the definition and possibilities possessed by people of Afrikan descent who chose to center themselves in their people: ancestors, present relatives, and those yet to be born into their lineage. This issue is especially important for psychologists, for it determines what Afrikan-centered therapeutic method best suits the needs of an individual in the realm of identity development.

Bynum (1999) echoes *The Kybalion* (Three Initiates, 1940) in insisting that what is now called psychology and what is now called metaphysics were, in Kemet, and are now the same entity. Kemet had created a better, more developed and balanced philosophy than is now available. Traditional societies such as that of the Bushmen of the Kalihari (Keeney, 1999, 2003) engage in healing by movement, rhythm, and surrendering control over their physical, emotional and mental selves. Western therapies, by contrast (using cognitive behavioral therapy as an example, since it dominates the landscape at present) require ever greater control, action but not necessarily movement, and no rhythm.

Healing in the West is a commodity with a class-based access and cost. In other words, one can obtain all the healing one can afford. In this way, healing functions like "justice" in the United States: on can get all the justice one can afford. This paradigm is not a holistic, shared way of doing things. Its goal is to function "normally" amidst insanity; its goal is not balance or reciprocity. The Afrikan concept of healing offers promise for closing the gaps in Afrikan psychological knowledge. The Kemites saw psychology and metaphysics, which is also called energy work, as combined into one vocation, called healing (Three Initiates, 1940). Afrikan peoples continue to use energy to transform, heal, and balance (Mutwa, 1996; Some, 1998, 2003; Suzar, 1999). Franklin and Moss (2000) note that Black communities in the United States included "healers" who used skills brought from Afrika along with folk cures to heal those in need (p. 110).

The quest for freedom and dignity for Afrikan Americans requires the best of our creativity. To give less than that would be, in effect, to give up (Akil, 1993). That is neither an object nor a goal. If it is true that life is a series of struggles, then Afrikan people are required to stand and face those struggles. Self-determination demands it; self-respect requires it; the children deserve it; and Afrikan ancestors expect nothing less of us (Fuller, 1984).

It is not by accident, but rather by intention, that the quotes that begin each chapter are from men. As this dissertation focuses on Afrikan male identity development, these quotes play a part in creating and recreating the Afrikan male. As Clarke (1999) notes, Afrikan males used to be secure enough to revel in the accomplishments of Afrikan women, supporting and encouraging them to go as far as their talents would allow. It is hoped that as Afrikan men continue to emerge into maleness, that Afrikan women will continue to emerge into femaleness, building and rebuilding Afrikan relationships, families, and nations. As it once was, so shall it be. Hotep. Ase'.

> We heal people, individuals. That's part of our work. But it isn't all. It isn't even the greater part of it . . . . Sometimes a whole people needs healing work. Not a tribe, not a nation. Tribes and nations are just signs that the whole is diseased. The healing work that cures a whole people is the highest work, far greater than the cure of single individuals. (Armah, 1978, p. 97)

Footnotes

[1] The terms "Africa" and "African" will be spelled "Afrika" and "Afrikan." This choice indicates a self-conscious Afrikan perspective is being used in the analysis (Amini, 1969; Barashango, 1983; Clarke, 1993; Kgositsile, 1971; Toure, 1973; Wilson, 1999).

[2] The terms "European" and "White" will be used interchangeably throughout, in response to their use by scholars and in the novel, *Invisible Man*.

[3] Carr (1997, pp. 285-286, note 2) distinguishes between "Afrocentric" and "African-centered" knowledge. Afrocentric knowledge relies on European epistemological premises for the production of knowledge, whereas Afrikan-centered knowledge operates with Afrika as its epistemological center.

[4] Smedley and Smedley (2005) argue persuasively that race continues to be a key ingredient in maintaining social stratification in the United States, despite recent advances in biology, anthropology and genetics, which argue that humans of all races are more alike than dissimilar, making the concept of race irrelevant.

[5] "Kemet" means "land of the Blacks" in the language of Mdw Ntr, the language written and spoken in what is commonly called ancient Egypt. The Greeks renamed the

language hieroglyphics, and renamed the country "Egypt" (Carruthers, 1995; Karenga, 2002).

[6]In a later example of this power, in 1493 the Pope issued a series of bulls or rules that divided the Afrikan continent and the "New World" for commercial exploitation between Portugal and Spain (Clarke, 1993).

[7]See Welsing's (1991) discussion of the symbolic nature of Europeans' desire for forbidden sex with persons of Afrikan descent, and their resultant projection of guilt and sin onto those persons who they have oppressed and/or suppressed.

[8] Azania is the Afrikan name of the country that is commonly referred to as South Africa(Biko, 1978).

[9]Kambon's term "anti-Afrikan/anti-Black" (1998, p. 171) will be substituted for the terms "White supremacy" and "European domination" where appropriate. Doing so keeps the discussion grounded with Afrika and Afrikans at the center of the analysis.

[10]Fu-Kiau (1991, p. 58) makes the distinction between a "client," which indicates a financial arrangement, and a "patient," which indicates a healing arrangement.

[11] Jacobs (1987) wrote *Incidents in the Life of a Slave Girl* under the pen name, Linda Brent. It was not until the late 1980s that research revealed that this story, which was presumed to be fictional, was, in fact, true. Jacobs was an escaped slave.

The book can still be found with the author listed either as Linda Brent or as Harriet Jacobs.

[12]See, for example, Swahili Name Book, 1971; Karenga, 1975; Karim, 1976; Asante, 1991; Stewart, 1996; Musere and Byakutaga, 2000; Adebayo, 2005; and www.namesite.com, which lists African names and their meanings.

[13] Washington was a spokesperson for persons of Afrikan descent in America in the late nineteenth and early twentieth centuries. In his address at the Atlanta Cotton Exposition in 1895, Washington suggested that persons of Afrikan descent stay in inferior positions with respect to persons of European descent after the Civil War. Washington preferred to build up economic stability for persons of Afrikan descent rather than obtain equality in all spheres of life in the United States (Washington, 1906).

[14]Ellison's parents named him Ralph Waldo Ellison after Ralph Waldo Emerson.

[15]As with many aspects of this novel, Brockway's name has symbolic meaning. "Brock" is another name for a badger, which, when taken as verb, describes the "Way" in which Brockway treats IM.verb, describes the "Way" in which Brockway treats IM.

[16] Frederick Douglass, who successfully escaped enslavement, abolitionist speaker and newspaper editor. As such, he represented the hope of all enslaved persons of Afrikan descent (Douglass, 1968).

[17]Horse-drawn Borden's milk wagons were manufactured until the 1940s, indicating that the novel is set in the 1930s or 1940s (*Bordon's Milk Delivery Wagon*, 2005).

[18]A second novel by Ellison, entitled *Juneteenth* (1999), was issued after his death.

[19]Transubstantive error occurs when the cultural and psychological norms of one group are applied to establish the meaning of the cultural and psychological functioning of another group (Azibo, 1996a).

[20]"Signifying" and "playing the dozens" are both verbal interchanges to test emotional strength. In the dozens, each contestant insults the other's relatives, and especially their mother (Major, 1970).

[21]A "mammy" is the nurturer of her master and his family. She does the cooking, cleaning, raising of the master's children, and cares deeply for the welfare of her oppressors (Boskin, 1973; Rhines, 1996).

[22]Griots and griottes orally transmitted the history of their village and clan from generation to generation (Hale, 1999). It was a griot in the Gambia, a small country in West Africa, who told Alex Haley the oral history of his family that is portrayed in his book, *Roots*. That history included how Haley's ancestor, Kunta Kinte, was kidnaped and sold into slavery (Haley, 1976).

[23]Carr (1997, p. 319, note 91) defines between a person who *locates* themselves as an Afrikan person as being without any criteria available to include or exclude them. On the other hand, to be Afrikan *oriented* requires observable, healthy cultural behavior, such as those enumerated by Akbar, Ani, Azibo, Goddard, Kambon, Nobles and other Afrikan-centered theorists.

[24]Bynum's (1999) collective Afrikan unconscious is defined as a central to all persons of Afrikan descent. Bynum goes further, saying that all of humanity possesses this collective Afrikan unconscious, since all people, so far as science can tell, are descended from ancestors of Afrikan descent.

References

Adebayo, B. (2005). *Dictionary of African names Volume I*. Bloomington, IN: AuthorHouse.

Ajamu, A. A. (2004). Rekh: Prelude to an intergenerational conversation about African psychological thought. In R. L. Jones, (Ed.), *Black psychology* (4th ed., pp. 221-242). Hampton, VA: Cobb & Henry.

Akbar, N. (1976). Rhythmic patterns in African personality. In L. M. King, V. J. Dixon, & W. W. Nobles (Eds.), *African philosophy: Assumptions & paradigms for research on Black persons* (pp. 175-189). Los Angeles: Fanon Center Publication.

Akbar, N. (1984). *Chains and images of psychological slavery*. Jersey City, NJ: New Mind Productions.

Akbar, N. (1985a). Nile valley origins of the science of the mind. In I. Van Sertima (Ed.), *Nile valley civilizations* (pp. 120-132). New Brunswick, NJ: Journal of African Civilizations.

Akbar, N. (1985). *The community of self*. Tallahassee, FL: Mind Productions.

Akbar, N. (1991). *Visions for Black men*. Tallahassee, FL: Mind Productions.

Akbar, N. (1994). *Light from ancient Africa*. Tallahassee, FL: Mind Productions.

Akbar, N. (2004). *Akbar papers in African psychology*. Tallahassee, FL: Mind Productions.

Akil. (1993). *From niggas to gods, part one*. St. Louis, MO: Eight Press. Alexander, C. (2005). *John Crow speaks: Earth teachings of the Jamaican elders*. Rhinebeck, NY: Monkfish Book.

Allen, N. R. (2005). *Blind "Afrocentricity" likely to lead Blacks astray*.

Retrieved May 11, 2005, from http://www.secularhumanism. org/library/aah/allen_1_1.html

Allen, R. L. (2001). *The concept of self: A study of Black identity and self esteem*. Detroit, MI: Wayne State University.

Allen, R. L., & Bagozzi, R. P. (2001). Consequences of the Black sense of self. *The Journal of Black Psychology, 27*(1), 3-28.

Allen-Meares, P., & Burman, S. (1995). The endangerment of African American men: An appeal for social work action. *Social Work, 40*(2), 268-274.

Amini, J. M. (1969). *An Afrikan frame of reference*. Chicago, IL: Institute of Positive Education.

Amuleru-Marshall, O. (1993). Political and economic implications of alcohol and other drugs in the African-American community. In L. L. Goddard (Ed.), *An African-centered model of prevention for African-American Youth at high risk* (pp. 23-33). Rockville, MD: U.S. Department of Health and Human Services.

Anderson, M. F. (1994). *Black English vernacular (from "ain't to yo mama"): The words politically correct Americans should know*. Highland City, FL: Rainbow Books.

Ani, M. (1994). *Yurugu: An African-centered critique of European thought and behavior*. Trenton, NJ: Africa World.

Armah, A. K. (1978). *The healers*. Popenguine, Senegal, West Africa: Per Ankh.

Asante, M. K. (1980). *Afrocentricity: The theory of social change*. Buffalo, NY: Amulefi.

Asante, M. K. (1991). *The book of African names*. Trenton, NJ: Africa World.

Austin, S. (2001). Race matters. *Radical Psychology, 2*(1), 1-4.

Azibo, D. A. A. (1988). Advances in Black/African personality theory. In , D. A. A. (Ed.), *Africentric Essays in African personality:*

Theory, practice, research (chap. 1). Unpublished manuscript.

Azibo, D. A. A. (1996a). African psychology in historical perspective and related commentary. In D. A. A. Azibo (Ed.), *African psychology in historical perspective & related commentary* (pp. 1-28). Trenton, NJ: Africa World Press.

Azibo, D. A. A. (1996b). Mental health defined Africentrically. In D. A. A. Azibo (Ed.), *African psychology in historical perspective & related commentary* (pp. 47-56). Trenton, NJ: Africa World Press.

Azibo, D. A. A., & Robinson, J. (2004). An empirically supported reconceptualization of African-U.S. racial identity development as an abnormal process. *Review of General Psychology, 8*(4), 249-264.

Baker, H. A., Jr. (1986). Creativity and commerce in the Trueblood episode. In H. Bloom (Ed.), *Ralph Ellison: Modern critical views* (pp. 113-128). New Haven, CT: Chelsea House.

Baldwin, J. (1948). *Go tell it on the mountain.* New York: Dell.

Baldwin, J. (1954). *Nobody knows my name.* New York: Dell.

Baldwin, J. (1972). *No name in the street.* New York: Dell.

Baldwin, J. A. (1986). African (Black) psychology: Issues and synthesis. *Journal of Black Studies, 16*(3), 235-249.

Baldwin, J. A. (1992). The role of Black psychologists in Black liberation. In A. K. H. Burlew, W. C. Banks, H. P. McAdoo, & D. A. A. Azibo (Eds.), *African American psychology: Theory, research, and practice* (pp. 48-57*).* Newbury Park, CA: Sage.

Banks, W. H. (1979). Psychohistory and the black psychologist. In W. D. Smith, K. H. Burlew, M. H. Mosley, & W. M. Whitney (Eds.), *Reflections on Black psychology* (pp. 33-40). Washington, DC: University Press of America.

Barashango, I. (1983). *Afrikan people and European holidays: A mental genocide: Books one and two*. Washington, DC: IVth Dynasty.

Bartky, S. L. (1990). *Femininity and domination: Studies in the phenomenology of domination*. New York: Routledge & Kegan Paul.

Baruti, M. K. B. (2005). *Mentacide and other essays*. Atlanta, GA: Akoben.

Bascom, W. (1992). *African folktales in the New World*. Bloomington, IN: Indiana University.

Baumbach, J. (1986). Nightmare of a native son. In H. Bloom (Ed.), *Ralph Ellison* (pp. 13-27). New York: Chelsea House.

Beatty, M. H. (1997). Maat: The cultural and intellectual allegiance of a concept. In J. H. Carruthers & L. C. Harris, (Eds). *African World History Project: The Preliminary Challenge* (pp. 211-244). Los Angeles: Association for the Study of Classical African Civilizations.

Beckson, K., & Ganz, A. (1989). *Literary terms: A dictionary* (3rd ed.). New York: Noonday.

Bell, D. (1992). *Faces at the bottom of the well: The permanence of racism*. New York: Basic.

ben-Jochannan, Y. A. A. (1963). *The Angola crisis and the rape of Africa*. New York: Alkebu-lan.

ben-Jochannan, Y. A. A. (1971). *Africa: Mother of western civilization*. New York: Alkebu-lan.

ben-Jochannan, Y. A. A. (1972). *The Black man's religion and extracts and comments from the holy Black Bible, V. II*. New York: Alkebu-lan.

ben-Jochannan, Y. A. A. (1981). *Black man of the Nile and his family*. New York: Alkebu-lan.

Bennett, L., Jr. (Ed.). (1971). *Ebony pictoral history of Black America, Volume III: Civil rights movement to Black revolution*. Chicago:

Johnson.

Bennett, L., Jr. (1988). *Before the Mayflower: A history of Black America* (6th ed.). New York: Penguin.

Bennett, S. B., & Nichols, W. W. (1974). Violence in Afro-American fiction: An Hypothesis. In J. Hersey (Ed.), *Ralph Ellison: A collection of critical essays* (pp. 171-175). Englewood Cliffs, NJ: Prentice-Hall.

Bianchi, F. T., Zea, M. C., Belgrave, F. Z., & Echeverry, J. J. (2002). Racial identity and self-esteem among Black Brazilian men: Race matters in Brazil too! *Cultural Diversity & Ethnic Minority Psychology, 8*(2), 157-169.

Biko, S. (1978). *I write what I like*. San Francisco: Harper.

Billingsley, A. (1968). *Black families in White America*. Englewood Cliffs, NJ: Prentice-Hall.

Billson, J. M. (1996). *Pathways to manhood: Young Black males struggle for identity*. New Brunswick, NJ: Transaction.

Blake, S. (1986). Ritual and rationalization: Black folklore in the works of Ralph Ellison. In H. Bloom (Ed.), *Ralph Ellison* (pp. 77-99). New York: Chelsea House.

Blyden, E. W. (1888/1994). *Christianity, Islam and the Negro race*. Baltimore: Black Classic.

Blyden, E. W. (1908/1994). *African life & customs*. Baltimore: Black Classic.

Bobo, J. (1995). *Black women as cultural leaders*. New York: Columbia.

Boker, P. A. (1996). *The grief taboo in American literature: Loss and prolonged adolescence in Twain, Melville, and Hemingway*. New York University.

Bone, R. (1958). *The Negro novel in America*. New Haven, CT: Yale University Press.

Bone, R. (1970). Ralph Ellison and the Use of Imagination. In. J. M. Reilly (Ed.), *Twentieth century interpretations of Invisible Man* (pp. 23-31). Englewood Cliffs, NJ: Prentice-Hall.

Borden's milk delivery wagon. (2005). Retrieved December 26, 2005, from http://images4.indianahistory.org/cdm4/item_viewer. php? CISOROOT'/po129&CISOPT

Boskin, J. (1973). Racial stereotyping and popular culture. In A. C. Gulliver (Ed.), *Proceedings of a Symposium on Black Images in Film, Stereotyping, and Self-perception as Viewed by Black Actresses* (pp. 6-8). Boston: Boston University Afro-American Studies Program.

Boskin, J. (1986). *Sambo: The rise & demise of an American jester.* New York: Oxford.

Boyd-Franklin, N. (1998). A multisystems approach to home and community based interventions with African American poor families. In R. L. Jones, (Ed.), *African American mental health: Theory, research, and intervention* (pp. 315-328). Hampton, VA: Cobb & Henry.

Boyd-Franklin, N., & Franklin, A.J. (2000). *Boys into men: Raising our African American teenage sons.* New York: Penguin.

Boykin, A. W. (1979). Black psychology and the research process: Keeping the baby but throwing out the bath water. In A. W. Boykin, A. J. Franklin, & J. F. Yates (Eds.), *Research directions of Black psychologists* (pp. 85-103). New York: Russell Sage.

Boykin, A. W., Franklin, A. J., & Yates, J. F. (1979). Work notes on empirical research in Black psychology. In A. W. Boykin, A. J. Franklin, & J. F. Yates (Eds.), *Research directions of Black psychologists* (pp. 3-19). New York: Russell Sage.

Bradford, R. (1931). *John Henry.* New York: The Literary Guild.

Brookins, C. C. (2004). Promoting ethnic identity development in

African American youth: The role of rites-of-passage. In R. L. Jones (Ed.), *Black* psychology (4th ed., pp. 509-539). Hampton, VA: Cobb & Henry.

Browder, A. T. (1992). *Nile valley contributions to civilization.* Washington, DC: The Institute of Karmic Guidance.

Budge, E. A. W. (1915). *The book of the dead: The hieroglyphic transcript of the Papyrus of ANI, the translation into English and an introduction by E. A. Wallis Budge, late keeper of the Egyptian and Assyrian antiquities in the British museum.* New York: Bell.

Bulhan, H. A. (1980). The revolutionary psychology of Frantz Fanon and some further notes on his theory of violence. *Fanon Center Journal, 1*(1), 51-71.

Bulhan, H. A. (1985). *Frantz Fanon and the psychology of oppression.* New York: Plenum Press.

Busby, M. (1991). *Ralph Ellison.* Boston: Twayne.

Butler, R. J. (2000). *The critical response to Ralph Ellison, volume 35.* Westport, CT: Greenwood.

Bynum, E. B. (1999). *The African unconscious: Roots of ancient mysticism and modern psychology.* New York: Teachers College.

Cabral, A. (1973). *Return to the source: Selected speeches of Amilcar Cabral.* New York: Monthly Review.

Callahan, J. F. (Ed.). (2004). *Ralph Ellison's Invisible Man: A casebook.* New York: Oxford.

Carmichael, S., & Hamilton, C. V. (1967). *Black power: The politics of liberation in America.* New York: Vintage.

Carr, G. E. K. (1997). The African-centered philosophy of history: An exploratory essay on the genealogy of foundationalist historical thought and African nationalist identity construction. In J. H. Carruthers & L. C. Harris, (Eds.), *African World History Project: The Preliminary Challenge* (pp. 285-320). Los Angeles:

Association for the Study of Classical African Civilizations.

Carruthers, J. H. (1984). *Essays in ancient Egyptian studies*. Los Angeles: University of Sankore.

Carruthers, J. H. (1995). *Mdu Ntr, divine speech: A historiciographical reflection of African deep thought from the time of pharoahs to the present*. London, England: Karnak House.

Carruthers, J. H. (1996). Science and oppression. In D.A. A. Azibo (Ed.), *African psychology in historical perspective & related commentary* (pp. 185-191). Trenton, NJ: Africa World.

Carruthers, J. H. (1999). *Intellectual warfare*. Chicago: Third World.

Caute, D. (1970). *Frantz Fanon*. New York: Viking.

Cesaire, A. (1969). *Return to my native land*. Baltimore: Penguin.

Chandler, W. B. (1999). *Ancient Future: The teachings and prophetic wisdom of the seven Hermetic laws of ancient Egypt*. Baltimore: Black Classic.

Chavous, T., Rivas, D., Green, L., & Heliare, L. (2002). Role of student background, perceptions of ethnic fit, and racial identification in the academic adjustment of African American students at a predominantly white university. *Journal of Black Psychology, 28*(3), 234-260.

Chinweizu (1975). *The west and the rest of us: White predators, Black slavers and the African elite*. New York: Vintage.

Chinweizu, Jemie, O., & Madubuike, I. (1983). *Toward the decolonization of African literature*. Washington, DC: Howard University.

Christian, B. (1969). Ralph Ellison: A critical study. In A. Gayle (Ed.), *Black expression: Essays by and about Black Americans in the creative arts* (pp. 353-365). New York: Weybright and Talley.

Citizens Commission on Human Rights. (1995). *Psychiatry's Betrayal*. Los Angeles: Citizens Commission on Human Rights.

Civil Rights Quotes. (2005). *Jesse Jackson: I am somebody Address to Operation Breadbasket rally, 1966.* Retrieved December 20, 2005, from http://www.historylearningsite.co.uk/civil%20rights%20quotes.htm

Clark C. X., McGee, D. P., Nobles, W., & Akbar, N. (1976). *Voodoo or IQ: An introduction to African psychology.* Chicago: Institute of Positive Education.

Clark, K. B. (1963). Black and White: The ghetto inside. In R. V. Guthrie (Ed.), *Being black: Psychological-sociological dilemmas* (pp. 66-72). San Francisco: Canfield.

Clark, K. B. (1965). *Dark Ghetto: Dilemmas of social power.* New York: Harper & Row.

Clarke, J. H. (Ed.). (1969). *Malcolm X: The man and his times.* Toronto, Canada: Collier-Macmillan.

Clarke, J. H. (1970). Ellison and his novel: The visible dimensions of *Invisible Man. Black World, 20*(2), 27-30.

Clarke, J. H. (Ed.). (1991). *New dimensions in African history: The London lectures of Dr. Yosef ben-Jochannan and Dr. John Henrik Clarke.* Trenton, NJ: Africa World Press.

Clarke, J. H.. (1993). *Christopher Columbus and the Afrikan holocaust: Slavery and the rise of European capitalism.* Brooklyn: A&B Publishers.

Clarke, J. H. (1994). *Who betrayed the African world revolution? and other speeches.* Chicago: Third World Press.

Clarke, J. H. (1999). *My life in search of Africa.* Chicago: Third World.

Clarke, J. H., & Garvey, A. J. (1974). *Marcus Garvey and the vision of Africa.* New York: Vintage.

Clegg, L. H., II & Ahmed, K. Y. (1999). *When Black men ruled the world: Part one: Egypt during the golden age.* Compton, CA: The

Clegg Series.

Coley, R. L. (2001). (In)visible men: Emerging research on low-income, unmarried and minority fathers. *American Psychologist, 56*(9), 743-753.

Collier, E. W. (1970). The nightmare truth of an invisible man. *Black World, 20*(2), 12-19.

Connor, M. E., & White, J. L. (Eds.). (2006). *Black fathers: An invisible presence in America*. Mahwah, NJ: Lawrence Erlbaum.

Cross, W. E., Jr. (1979). The Negro-to-Black conversion experience: An empirical analysis. In A. W. Boykin, A. J. Franklin, & J. F. Yates (Eds.), *Research directions of Black psychologists* (pp. 107-130). New York: Russell Sage.

Cross, W. E., Jr. (1980). Models of psychological nigrescence: A literature review. In R. L. Jones (Ed.), *Black Psychology* (2nd ed., pp. 81-98). New York: Harper & Row.

Cross, W. E., Jr. (1991). *Shades of Black: Diversity in African-American identity*. Philadelphia: Temple University.

Cross, W. E., Jr., Parham, T. A., & Helms, J. E. (1991). The stages of Black identity development: Nigrescence models. In R. L. Jones (Ed.), *Black psychology* (3rd ed., pp. 319-338). Berkeley, CA: Cobb & Henry. Cruse, H. (1967). *The crisis of the Negro intellectual*. New York: New York Review Books.

Davidson, B. (1964). *Which way Africa? The search for a new society* (3rd ed.). Baltimore: Penguin.

Davis, C. T. (1986). The mixed heritage of the modern Black novel. In H. Bloom (Ed.), *Ralph Ellison: Modern critical views* (pp. 101-111). New Haven, CT: Chelsea House.

Davis, F. J. (1991). *Who is Black? One nation's definition*. University Park, PA: The Pennsylvania State University Press.

de Quincey, C. (2005). *Radical knowing: Understanding consciousness through relationship.* Rochester, VT: Bear.

Delany, M. R. (1879/1991). *Principia of ethnology: The origin of races and color, with an archeological compendium of Ethiopian and Egyptian civilization, from years of careful examination and enquiry.* Baltimore: Black Classic.

Dennard, D. (1998). Application of the Azibo nosology in clinical practice with Black clients: A case study. *Journal of Black Psychology, 24*(2), 182-195.

Dillard, J. L. (1972). *Black English: Its history and usage in the United States.* New York: Vintage.

Diop, C. A. (1974). *The African origin of civilization: Myth or reality.* Chicago: Lawrence Hill.

Diop, C. A. (1980). *The genetic parentage of pharoanic Egyptian and the languages of Black Africa.* Unpublished manuscript.

Diop, C. A. (1987). *Precolonial Black Africa: A comparative study of the political and social systems of Europe and Black Africa, from antiquity to the formation of modern states.* Brooklyn, NY: Lawrence Hill.

Diop, C. A. (1990). Origin of the ancient Egyptians. In G. Mokhtar (Ed.), *General history of Africa II: Ancient civilizations of Africa* (pp. 15-61). Berkeley, CA: University of California.

Diop, C. A. (1991). *Civilization or barbarism: An authentic anthropology.* Brooklyn, NY: Lawrence Hill.

Dismukes, G. (1995). *Afrikan alchemy.* Nashville, TN: One Horn Press.

Dixon, V. J. (1976). World views and research methodology. In L. M. King, V. J. Dixon, & W. W. Nobles (Eds.), *African philosophy: Assumptions & paradigms for research on Black persons* (pp. 51-102). Los Angeles: Fanon Center Publication.

Dixon, V. J., & King, L. M. (1980). Frantz Fanon: A biographical note: Theman; the mind, the message. *Fanon Center Journal, 1*(1), 73-79.

Dostoyevsky, F. M. (1864/1948). Notes from underground. In C. Neider (Ed.). *Short novels of the masters* (pp. 125-219). New York: Holt, Rinehart & Winston.

Douglass, F. (1968). *My bondage and freedom*. New York: Arno Press.

Dubey, M. (2003). Postmodernism as postnationalism? Racial representation in U.S. Black cultural studies. *The Black Scholar, 33(1)*, 2-18.

DuBois, W. E. B. (1896/1969). *The suppression of the African slave trade to the United States of America: 1638-1870*. Baton Rouge, LA: Louisiana State University.

DuBois, W. E. B. (1946/1965). *The world and Africa: An inquiry into the part which Africa has played in world history*. New York: International.

DuBois, W. E. B. (1961). *The souls of Black folk*. New York: Dodd, Mead & Company.

DuBois, W. E. B. (1978). *On the importance of Africa in world history*. Harlem, NY: Black Liberation Press.

Dunbar, P. L. (1913). *The complete poems of Paul Laurence Dunbar*. New York: Dodd, Mead.

Dyson, M. E. (2004). *Mercy, mercy me: The art, loves & demons of Marvin Gaye*. New York: Civitas.

Edel, L. (1964). *The modern psychological novel*. New York: Grosset & Dunlap.

Edel, L. (1982). *Stuff of sleep and dreams: Experiments in literary psychology*. New York: Harper & Row.

Edwards, K. L. (1999). African American definitions of self and psychological health. In R. L. Jones (Ed.), *Advances in African American psychology* (pp. 287-312). Hampton, VA: Cobb &

Henry.

Eichelberger, J. (1999). *Prophets of recognition: Ideology and the individual in novels by Ralph Ellison, Toni Morrison, Saul Bellow, and Eudora Welty.* Baton Rouge, LA: Louisiana State University.

Elkins, D. N. (1998). *Beyond religion: A personal program for building a spiritual life outside the walls of traditional religion.* Wheaton, IL: Quest Books.

Ellison, R. (1947). *Invisible man.* New York: Vintage.

Ellison, R. (1966). *Shadow and act.* New York: Signet Books.

Ellison, R. (1999). *Juneteenth.* New York: Random House.

Ephirim-Donkor, A. (1997). *African spirituality: On becoming ancestors.* Trenton, NJ: Africa World Press.

Fairchild, H. H. (2004). Solving the "acid tests" of African psychology. In R. L. Jones (Ed.). *Black psychology* (4th ed., pp. 213-219). Hampton, VA: Cobb & Henry.

Fanon, F. (1963). *The wretched of the earth.* New York: Grove.

Fanon, F. (1965). *A dying colonialism.* New York: Grove.

Fanon, F. (1967a). *Black skin, white masks.* New York: Grove.

Fanon, F. (1967b). *Toward the African revolution.* New York: Monthly Review.

Fetterman, D. M., Kaftarian, S. J., & Wandersman, A. (Eds). (1996). *Empowerment evaluation: Knowledge and tools for self-assessment & accountability.* Thousand Oaks, CA: Sage.

Ford, N. A. (1970). The ambivalence of Ralph Ellison. *Black World, 20*(2), 5-9.

Forrest, L. (2000). Luminosity from the Lower Frequencies. In R. J. Butler (Ed.), *The critical response to Ralph Ellison* (pp. 61-71). Westport, CT: Greenwood.

Franklin, A. J. (1992). Therapy with African American men. *Journal of*

Contemporary Human Services, 73, 350-355.

Franklin, A. J. (1998). Invisibility syndrome in psychotherapy with African American males. In R. L. Jones (Ed.), *African American mental health: Theory, research, and intervention* (pp. 395-411). Hampton, VA: Cobb & Henry.

Franklin, A. J. (2002). *From brotherhood to manhood: How Black men rescue their relationships and dreams from the invisibility syndrome.* New York: John Wiley.

Franklin, J. H., & Moss, A. A., Jr. (2000). *From slavery to freedom* (8th ed.). New York: McGraw-Hill.

Fredrickson, G. M. (1981). *White supremacy: A comparative study in American & South African history.* New York: Oxford.

Fredrickson, G. M. (1988). *The arrogance of race: Historical perspectives on slavery, racism and social inequality.* Hanover, NH: Wesleyan University.

Freud, S. (1946). *Totem and taboo: resemblances between the psychic lives of savages and neurotics.* New York: Vintage.

Fu-Kiau, K. K. B. (1991). *Self-healing power and therapy: Old teachings from Africa.* Baltimore: Imprint Editions/Black Classic.

Fu-Kiau, K. K. B., & Lukondo-Wamba, A. M. (1988). *Kindezi: the Kongo art of babysitting.* Baltimore: Imprint Editions/Black Classic.

Fuller, N, Jr. (1984). *The united independent compensatory code/system/concept: a textbook/workbook for thought, speech and/or action for victims of racism (white supremacy).* Washington, DC: Author.

Gadalla, M. (1999). *Exiled Egyptians: The heart of Africa.* Greensboro, NC: Tehuti Research Foundation.

Gadalla, M. (2001). *Egyptian divinities: The all who are the One.* Greensboro, NC: Tehuti Research Foundation.

Gardiner, S. A. (1927). *Egyptian grammar: Being an introduction to the study of hieroglyphs*. Oxford, England: Griffith Institute.

Gates, H. L., Jr., & Appiah, K. A., (Eds.). (1993). *Richard Wright: Critical perspective past and present*. New York: Amistad.

Gates, H. L., Jr., McKay, N. Y., Andrews, W. L., Baker, H. A., Jr., Christian, B. T., Foster, F. S., et al. (Eds.). (2004) *The Norton anthology of African American literature* (2nd ed.). New York: W.W. Norton.

Gayle, A., Jr. (1969). Perhaps not so soon one morning. In A. Gayle (Ed.). *Black expression: Essays by and about Black Americans in the creative arts.* (pp. 280-288). New York: Weybright & Talley.

Gayle, A., Jr. (1970a). *The black situation*. New York: Dell.

Gayle, A., Jr. (1970b). The function of Black literature at the present time. In R. Smith, & S. L. Jones (Eds.), *The Prentice Hall anthology of African American literature.* (pp. 977-984). Upper Saddle River, NJ: Prentice Hall.

Gayle, A., Jr. (1972). *The Black aesthetic*. Garden City, NY: Anchor.

Gibson, D. B. (1971). Ralph Ellison and James Baldwin. In G. A. Panichas (Ed.), *The politics of twentieth-century novelists* (307-320). New York: Hawthorn Books.

Gibson, D. B. (1981). *The politics of literary expression: A study of major Black writers*. Wesport, CT: Greenwood Press.

Gillem, A. R., Cohn, L. R., & Throne, C. (2001). Black identity in biracial Black/White people: A comparison of Jacqueline who refuses to be exclusively Black and Adolphus who wishes he were. *Cultural Diversity and Ethnic Minority Psychology, 7*(2), 182-196.

Ginsberg, E. K. (Ed.). (1996). *Passing and the fictions of identity*. Durham, NC: Duke University.

Goddard, L. L. (1993). Natural resistors in AOD abuse prevention in the African-American family. In L. L. Goddard (Ed.), *An African-centered model of prevention for African-American youth at high risk* (pp. 73-77). Rockville, MD: U.S. Department of Health and Human Services.

Gordon, L. R., (Ed). (1995). *Fanon and the crisis of European man: An essay on philosophy and human* sciences. New York: Routledge.

Gordon, L. R., (Ed). (1997). *Existence in Black: An anthology of Black existential philosophy*. New York: Routledge.

Gordon, L. R. (2000). *Existentia Africana: Understanding Africana existential thought*. New York: Routledge.

Graham, M. J. (1999). The African-centered worldview: Toward a paradigm for social work. *Journal of Black Studies, 30*(1), 103-122.

Gray, V. B. (1978). *Invisible Man's literary heritage: Beneto Cereno and Moby Dick*. Amsterdam, Netherlands: Editions Rodopi N.V.

Greene, J. L. (1996). *Blacks in eden: The African American novel's first century*. Charlottesville, VA: University of Virginia.

Gregory, W. H., & Harper, K. W. (2001). The Ntu approach to health and healing. *Journal of Black Psychology, 27*(3), 304-320.

Grier, W. H., & Cobbs, P. M. (1968). *Black rage*. New York: Bantam.

Grills, C. T. (2004a). African psychology. In R. L. Jones (Ed.), *Black psychology*, (4th ed., pp. 171-208). Hampton, VA: Cobb & Henry.

Grills, C. T. (2004b). African psychology: Rejoinder: The search for authenticity and legitimacy. In R. L. Jones (Ed.), *Black psychology*, (4th ed., pp. 243-260). Hampton, VA: Cobb & Henry.

Grills, C. T. (2004c). To be an African psychologist: A pilgrimage to iwapele. *Psych Discourse, 36*(2), 5-7.

Grills, C., & Longshore, D. (1996). Africentrism: Psychometric analyses of a self-report measure. *Journal of Black Psychology, 22*(1), 86-106.

Guthrie, R. V. (1976). *Even the rat was white: A historical view of psychology.* New York: Harper & Row.

Hale, T. A. (1999). *Griots and griottes: Masters of words and music.* Bloomington, IN: Indiana University.

Haley, A. (1976). *Roots: The saga of an American family.* Garden City: Doubleday.

Hall, P. A. (1999). *In the vineyard: Working in African American studies.* Knoxville, TN: The University of Tennessee.

Hall, W. S., Cross, W. E., Jr., & Freedle, R. (1972). Stages in the development of Black awareness: An exploratory investigation. In R. L. Jones (Ed.), *Black Psychology* (1st ed., pp. 156-165). New York: Harper & Row.

Halsell, G. (1969). *Soul sister.* Greenwich, CT: Fawcett.

Hamilton, V. (1985). *The people could fly: American Black folktales.* New York: Alfred A. Knopf.

Harding, V. (1981). *There is a river: The Black struggle for freedom in America.* New York: Harcourt Brace Jovanovich.

Hare, N., & Hare, J. (1985). *Bringing the Black boy to manhood: The passage.* San Francisco: The Black Think Tank.

Harper-Browne, C. (1996). Toward curriculum development in Black psychology. In D. A. A. Azibo (Ed.), *African psychology in historical perspective & related commentary* (pp. 235-245). Trenton, NJ: Africa World Press.

Hawthorn, J. (1992). *A glossary of contemporary literary theory.* New York: Edward Arnold.

Helms, J. E. (1993). An overview of Black racial identity theory. In J. E. Helms(Ed.), *Black and white racial identity: Theory, research and practice* (pp. 9-32). Westport, CT: Praeger.

Helms, J. E. (1994).The conceptualization of racial identity and other "racial"constructs. In E. J. Trickett, R. J. Watts, & D. Birman (Eds.), *Human diversity: Perspectives on people in context* (pp. 285-311). San Francisco: Jossey-Bass.

Herbert, L. C. (2002). Can't you see what I'm saying? Making expressiveconduct a crime in high-crime areas. *Georgetown Journal on Poverty Law & Policy, 9*(1), 135-166.

Herbert, L. C. (2003). Bete noire: How race-based policing threatens national security. *Michigan Journal of Race & Law, 9*(1), 149-214.

Herbert, L. C. (2005). Plantation lullabies: How Fourth Amendment policing violates the Fourteenth Amendment right of African Americans to parent. *St. John's Journal of Legal Commentary, 19*(2), 197-235.

Herrnstein, R. J., & Murray, C. (1994). *The bell curve: Intelligence and class structure in American life*. New York: Free Press.

Herskovits, M. J. (1958). *The myth of the Negro past*. Boston: Beacon.

Hill, R. A., & Blair, B. (Eds.). (1987). *Marcus Garvey: Life and lessons.* Los Angeles: University of California.

Hilliard, III, A. G. (Ed.). (1991). *Testing African American students.* Morristown, NJ: Aaron Press.

Holdstock, T. L. (2000). *Re-examining psychology: Critical perspectives and African insights.* Philadelphia: Routledge.

Holdstock, T. L. (2002). Postmodern psychology and Africa. *American Psychologist. 57*(6/7), 460-461.

Holland, N. H., Homan, S., & Paris, B. J. (Eds.). (1989). *Shakespeare's personality.* Berkeley, CA: University of California.

Holloway, J. E. (2006). *African-American names.* Retrieved on March 3, 2006, from http://www.slaveryinamerica.org/history/hs_es_names.htm

Horowitz, F. R. (2000). Ralph Ellison's modern version of Brer Bear and Brer Rabbit in *Invisible Man.* In R. J. Butler (Ed.), *The critical response to Ralph Ellison* (pp. 45-49). Westport, CT: Greenwood.

Howe, I. (1974). Black boys and native sons. In J. Hersey (Ed.), *Ralph Ellison: A collection of critical essays* (pp. 36-38). Englewood Cliffs, NJ: Prentice-Hall.

Howard-Pitney, D. (2004). *Martin Luther King Jr., Malcolm X, and the Civil Rights struggle of the 1950s and 1960s: A brief history with documents.* New York: Beford/St. Martins.

Hurston, Z. N. (1937). *Their eyes were watching God.* Greenwich, CT: Fawcett.

Hyman, S. E. (1974). Ralph Ellison in Our Time. In J. Hersey (Ed.), *Ralph Ellison: A collection of critical essays* (pp. 39-42). Englewood Cliffs, NJ: Prentice-Hall.

Jackson, E. M. (1970). The American Negro and the image of the absurd. In J. M. Reilly (Ed.), *Twentieth century interpretations of Invisible Man* (pp. 64-72). Englewood Cliffs, NJ: Prentice-Hall.

Jackson, J. G. (1970). *Introduction to African civilizations.* New York: Citadel.

Jackson, L. P. (2002). *Ralph Ellison: Emergence of genius.* New York: John Wiley.

Jackson, R. L., II, & Stewart, J. B. (2002). *Negotiation of African American identities in rural America: A cultural contracts approach.* Retrieved on November 24, 2002, from http://www.marshall.edu/jrcp/ JacksonStewart.html

Jacobs, H. A. (1987). *Incidents in the life of a slave girl.* Cambridge, MA:Harvard.

James, G. G. M. (1976). *Stolen legacy.* San Francisco: Julian

Richardson.

Jenkins, A. H. (1995). *Turning corners: The psychology of African Americans*. Needham Heights, MA: Allyn & Bacon.

Jenkins, A. H. (1998). The self-concept in the psychology of African Americans. In R. L. Jones (Ed.), *African American mental health: Theory, research, and intervention* (pp. 167-185). Hampton, VA: Cobb & Henry.

Jenkins, A. H. (2004). A humanistic approach to African American psychology. In R. L. Jones (Ed.), *Black Psychology* (4th ed., pp. 135-155). Hampton, VA: Cobb & Henry.

Jensen, R. (2005). *The heart of Whiteness: Confronting race, racism, and White privilege*. San Francisco: City Lights.

Johnson, J. W. (1912/1960). *The autobiography of an ex-coloured man*. New York: Hill & Wang.

Jones, J. M. (1979). Conceptual and strategic issues in the relationship of Black psychology to American social science. In A. W. Boykin, A. J. Franklin, & J. F. Yates (Eds.), *Research directions of Black psychologists* (pp. 390-432). New York: Russell Sage.

Jones, J. M. (1991). Racism: A cultural analysis of the problem. In R. L. Jones, (Ed.), *Black psychology* (3rd ed., pp. 609-635). Berkeley, CA: Cobb & Henry.

Jones, L. (1963). *Blues people: The Negro experience in White American and the music that developed from it*. New York: William Morrow.

Jones, L., & Neal, L (Eds.) (1968). *Black fire: An anthology of Afro-American writing*. New York: William Morrow.

Jones, R. L. (Ed.). (1998). *African American identity development*. Hampton, VA: Cobb & Henry.

Jubal, J. A. (1991). *Black truth*. Long Beach, CA: Black Truth Enterprises.

Kaiser, E. (1970). A critical look at Ellison's fiction and at social and literary criticism by and about the author. *Black World, 20*(2), 53-97.

Kambon, K. K. K. (1992). *The African personality in America: An African-entered framework.* Tallahassee, FL: Nubian Nations.

Kambon, K. K. K. (1998). *African/Black psychology in the American context: An African-centered approach.* Tallahassee, FL: Nubian Nations.

Kambon, K. K. K. (2003). *Cultural misorientation: The greatest threat to the survival of the Black race in the 21st century.* Tallahassee, FL: Nubian Nations.

Kane, C. H. (1969). *Ambiguous adventure.* New York: Collier.

Karenga, M. R. (1967). *The quotable Karenga.* Los Angeles: Us Organization.

Karenga, M. R. (1975). *Names for a new people.* San Diego: Kawaida Publications.

Karenga, M. (1980). *Kawaida theory: An introductory outline.* Inglewood, CA: Kawaida Publications.

Karenga, M. (1993). *Introduction to Black studies* (2nd ed.). Los Angeles: University of Sankore Press.

Karenga, M. (2002). *Introduction to Black studies* (3rd ed.), Los Angeles: University of Sankore Press.

Karenga, M., & Carruthers, J. H. (1986). *Kemet and the African worldview: Research, rescue and restoration.* Los Angeles: University of Sankore.

Karim, Y. A. (1976). *Afrikan names.* Drewyville, VA: Afram Press.

Keeney, B. (Ed.) (1999). *Kalahari Bushmen healers.* Philadelphia: Ringing Rocks.

Keeney, B. (Ed.) (2003). *Ropes to God: Experiencing the Bushman spiritual universe.* Philadelphia: Ringing Rocks.

Kelley, N. (2004). *The head Negro in charge syndrome: The dead end of Black politics*. New York: Nation Books.

Kellner, B. (Ed.). (1979). *"Keep A-Inchin' Along": Selected writings of Carl Van Vechten about Black art and letters*. Westport, CT: Greenwood.

Kellogg, P. U. (Ed.). (1925). Harlem: Mecca of the new Negro. *Survey Graphic, VI*(6).

Kent, G. (1972). *Blackness and the adventure of western culture*. Chicago: Third World.

Kent, G. E. (1974). Ralph Ellison and the Afro-American folk and cultural tradition. In J. Hersey (Ed.), *Ralph Ellison: A collection of critical essays* (pp. 160-170). Englewood Cliffs, NJ: Prentice-Hall.

Kgositsile, K. (1971). *My name is Afrika*. Garden City, NY: Anchor.

King, L. M. (1980). Models of meaning in mental health: Model eight-- the transformation of the oppressed. *Fanon Center Journal, 1*(1), 29-49.

King, L. M., Dixon, V. J., & Nobles, W. W. (Eds.). (1976). *African philosophy: Assumptions & paradigms for research on Black persons* (pp. 51-102). Los Angeles: Fanon Center Publication.

King, M. L. (1957, August 1). The power of peaceful persuasion [CD No.BB0115]. Berkeley, CA: Pacifica Radio Archives.

King, R. (1994). *African origin of biological psychiatry*. Chicago, IL: Lushena Books.

Kitwana, B. (2003). *The hip hop generation: Young Blacks and the crisis in African American culture*. New York: Basic Civitas.

Kunjufu, J. (1985). *Countering the conspiracy to destroy Black boys, Vol. I*. Chicago: African American Images.

Kunjufu, J. (1986). *Countering the conspiracy to destroy Black boys, Vol. II*. Chicago: African American Images.

Kwate, N. O. A. (2003). Cross-validation of the Africentrism scale. *Journal of Black Psychology, 29*(3), 308-324.

Lamming, G. (1954). *In the castle of my skin.* New York: McGraw-Hill.

Lane, C., (Ed.). (1998). *The psychoanalysis of race.* New York: Columbia.

Latif, S. A., & Latif, N. (1994). *Slavery: The African American psychic trauma.* Chicago: Latif Communications Group.

Laye, C. (1954). *The dark child: The autobiography of an African boy.* New York: Farrar, Straus & Giroux.

Lee, C. C., & Bailey, D. F. (1998a). Counseling African Americans: Part 1 - Counseling African American men. *ACAeNews, 1*(4), 1-18.

Lee, C. C., & Bailey, D. F. (1998b). Counseling African Americans: Part 3 - Counseling African American male youth. *ACAeNews, 1*(5), 1-10.

Lehan, R. D. (1970). The Strange Silence of Ralph Ellison. In. J. M. Reilly (Ed.), *Twentieth century interpretations of Invisible Man* (pp. 106-110). Englewood Cliffs, NJ: Prentice-Hall.

Lester, J. (1969). *Black folktales.* New York: Grove.

Lewis, M. C. (1988). *Herstory: Black female rites of passage.* Chicago: African American Images.

Feldstein, R., & Mele, K. (Eds.). *Literature and psychology.* Providence, RI: Rhode Island College.

Lock, H. (1994). *A case of mis-taken identity: Detective undercurrents in recent African American fiction.* New York: Peter Lang.

Locke, A. (Ed.). (1975). *The new Negro.* New York: Atheneum.

Long, L. C. (1993). An Afrocentric intervention strategy. In L. L. Goddard (Ed.), *An African-centered model of prevention for African-American youth at high risk* (pp. 87- 92). Rockville,

MD: U.S. Department of Health and Human Services.

Long, N. (2005, December). *On the scene: Hip-hop's reconstruction of the gangster rap identity*. Retrieved on December 19, 2005, from http//www.indie-music.com/modules.php?name'News& file'article&sid'4504

Lott, B. (2002). Cognitive and behavioral distancing from the poor. *American Psychologist, 57*(2), 100-110.

Lucas, F. L. (1957). *Literature and psychology*. Ann Arbor, MI: University of Michigan Press.

Madhubuti, H. R. (1990). *Black men: Obsolete, single, dangerous? The Afrikan American family in transition: Essays in discovery, solution and hope*. Chicago: Third World.

Major, C. (1968). A Black criterion. In A. Chapman (Ed.), *Black voices: An anthology of African-American literature* (pp. 698-699). New York: Signet.

Major, C. (1970). *Dictionary of Afro-American slang*. New York: International Publishers.

Majors, R., & Billson, J. M. (1992). *Cool pose: The dilemmas of Black manhood in America*. New York: Touchstone.

Marks, B., Settles, I. H., Cooke, D. Y., Morgan, L., & Sellers, R. M. (2004). African American racial identity: A review of contemporary models and measures. In R. L. Jones (Ed.), *Black psychology* (4th ed., pp. 383-404). Hampton, VA: Cobb & Henry.

Martin, T. (1976). *Race first: The ideological and organizational struggles of Marcus Garvey and the Universal Negro Improvement Association*. Dover, MA: The Majority Press.

Mason, C. (1970). Ralph Ellison and the underground man. *Black World, 20*(2), 20-26.

Mazrui, A. A. (1986). *The Africans*. London, England: BBC

Publications.

Mbalia, D. D. (1995). *John Edgar Wideman: Reclaiming the African personality*. London, England: Associated University Presses.

Mbiti, J. S. (1970). *African religions and philosophy*. Garden City, NY: Anchor.

Melhem, D. H. (1987). *Gwendolyn Brooks: Poetry and the heroic voice*. Lexington, KY: The University Press of Kentucky.

Merriam-Webster's Collegiate Dictionary, (10th ed.). (2001). Springfield, MA: Merriam-Webster.

Moreland, R. C. (1999). *Learning from difference: Teaching Morrison, Twain, Ellison, and Eliot*. Columbus, OH: Ohio State University.

Moore, M. (2001). *Stupid White men . . . and other sorry excuses for the state of the nation*. New York: HarperCollins.

Morris, M. (1993). The complex nature of prevention in the African-American community: The problem of conceptualization. In L. L. Goddard (Ed.), *An African-centered model of prevention for African-American Youth at high risk* (pp. 59-71). Rockville, MD: U.S. Department of Health and Human Services.

Morrison, T. (1992). *Playing in the dark: Whiteness and the literary imagination*. Cambridge, MA: Harvard University.

Moses, W. J. (1989). The novel and its Afro-American, American, and European traditions. In S. R. Parr, & P. Savery (Eds.), *Approaches to Teaching Ellison's Invisible Man* (pp. 58-64). New York: The Modern Language Association of America.

Muhammad, E. (1965). *Message to the Blackman in America*. Chicago: Muhammad's Temple No. 2.

Murray, A., & Callahan, J.F. (Eds.). (2000). *Trading twelves: The selected letters of Ralph Ellison and Albert Murray*. New York: Modern Library.

Murray, B. (2002, January). Psychology bolsters the world's fight against racism. *Monitor on Psychology*, 52-53.

Musere, J., & Byakutaga, S. C. (2000). *African names and naming.* Los Angeles: Ariko.

Mutwa, V. C. (1996). *Zulu shaman: Dreams, prophecies, and mysteries.* Rochester, VT: Destiny Books.

Myers, L. J. (1991). Expanding the psychology of knowledge optimally: The importance of world view revisited. In R. L. Jones (Ed.), *Black Psychology* (4th ed. pp. 15-28). Berkeley, CA: Cobb & Henry.

Myers, L. J. (1993). *Understanding an Afrocentric worldview: Introduction to an optimal psychology.* Dubuque, IA: Kendall/Hunt.

Myers, L. J. (1999). Therapeutic processes for health and wholeness in the 21st century: Belief systems analysis and the paradigm shift. In R. L. Jones (Ed.), *Advances in African American psychology*, (pp. 313-355), Hampton, VA: Cobb & Henry.

Myers, L. J., Abdullah, S., & Leary, G. (2000). Conducting research with persons of African descent. In *Guidelines for Research in Ethnic Minority Communities.* Washington, D.C.: Council of National Psychological Associations for the Advancement of Ethnic Minority Interests (pp. 5-8). Washington, D.C.: American Psychological Association.

Myers, L. J., & Haggins, K. L. (1998). Optimal theory and identity development: Beyond the Cross Model. In R. L. Jones (Ed.), *African American Identity Development* (pp. 255-274). Hampton, VA: Cobb & Henry.

Nadel, A. (1988). *Ralph Ellison and the American canon: Invisible criticism.* Iowa City, IA: University of Iowa.

NameSite.com: *African names and meanings.* Retrieved on April 10, 2006, from http://www.namesite.com/

Natoli, J. (Ed.). (1984). *Psychological perspectives on literature: Freudian dissidents and non- Freudians: a casebook*. Hamden, CT: Archon.

Neal, L. (1968). An afterward: And Shine swam on. In L. Jones & L. Neal (Eds.), *Black fire: An anthology of Afro-American writing* (pp. 637-656). New York: William Morrow.

Neal, L. (1970). Ellison' zoot suit. *Black World, 20*(2), 31-52.

Newton, H. P. (1973). *Revolutionary suicide*. New York: Harcourt Brace Jovanovich.

Nichols, E. J. (1974). *The philosophical aspects of cultural difference*. Washington, DC: Nichols and Associates.

Nichols, W. (1970). Ralph Ellison's Black American scholar. *Phylon, 31*(1), 70-75.

Nkrumah, K. (1965). *Neo-colonialism: The last stage of imperialism*. New York: International.

Nkrumah, K. (1970). *Consciencism: Philosophy and ideology for decolonization*. New York: Modern Reader.

Nobles, W. W. (1972). African philosophy: Foundations for Black psychology. In R. L. Jones (Ed.), *Black psychology* (1st ed., pp. 18-32). New York: Harper & Row.

Nobles, W. W. (1976). African science: The consciousness of self. In L. M. King, V. J. Dixon, & W. W. Nobles (Eds.), *African philosophy: Assumptions & paradigms for research on Black persons*, (pp. 163-174). Los Angeles: Fanon Center Publication.

Nobles, W. W. (1980). Extended self: Rethinking the so-called Negro self-concept. In R. L. Jones (Ed.), *Black psychology* (2nd ed., pp. 99-105). New York: Harper & Row.

Nobles, W. W. (1986). *African psychology: Toward its reclamation, reascension & revitalization*. Oakland, CA: Black Family Institute.

Nobles, W. W. (1998). To be African or not to be: The question of identity or authenticity- some preliminary thoughts. In R. L. Jones (Ed.). *African American identity development* (pp. 185-206). Hampton, VA: Cobb & Henry.

Nobles, W. W., & Goddard, L. L. (1993). An African-centered model of prevention for African-American youth at high risk. In L. L. Goddard (Ed.), *An African-centered model of prevention for African-American youth at high risk* (pp. 115-129). Rockville, MD: U.S. Department of Health and Human Services.

Nobles, W. W., & Goddard, L. L. (1996). *An introduction to the African centered behavior change model for the prevention of HIV/STDs.* Washington, DC: The Association of Black Psychologists.

Obadele, I. A. (1975). *Foundations of the Black nation.* Detroit, MI: House of Songhay.

Obadele, I. A. (1997). *A brief history of Black struggle in America.* Baton Rouge, LA: The Malcolm Generation.

Obadele, I. A. (2003). The enemy's psychological assaults on the armed Black independence movement. In D. A. A. Azibo (Ed.), *African-Centered Psychology: Cultural-Focusing for Multicultural Competence* (pp. 221-239). Durham, NC: Carolina Academic Press.

Obasi, E. M., Prince, K. J., Bolden, M., Richardson, C., & Walker, R. L. (2005). Akoben: A battle cry for ABPsi's re-awakening. *Psych Discourse. 39*(6), 4-9.

Obenga, T. (1997). The genetic linguistic relationship between Egyptian (ancient Egyptian and Coptic) and modern Negro-African languages. In C. A. Diop, J. Leclant, T. Obenga, & J. Vercoutter, *The peopling of ancient Egypt & the deciphering of the meroitic script* (pp. 65-71). London, England: Karnak.

Obenga, T. (2004). *African philosophy: The pharaonic period: 2780-330*

BC. Popenguine, Senegal, West Africa: Per Ankh.

Olomenji. (1996). Menticide, genocide, and national vision: The crossroads for the Blacks of America. In D. A. A. Azibo (Ed.), *African psychology in historical perspective & related commentary,* (pp. 71-82). Trenton, NJ: Africa World Press.

O'Meally, R. G. (1980). *The craft of Ralph Ellison.* Cambridge, MA: Harvard.

Onyefulu, I. (2004). *Welcome, Dede: An African naming ceremony.* London, England: Francis Lincoln.

Oshodi, J. E. (2005). *Back then and right now in the history of psychology: A history of human psychology in African perspectives for the new millennium.* Bloomington, IN: AuthorHouse.

Padmore, G. (1972). *Pan-Africanism or communism.* New York: Anchor.

Parker, G. W. (1918/1981). *The children of the sun.* Baltimore: Black Classic.

Peery, N. (1975). *The Negro national colonial question.* Chicago: Workers.

Petrie, F. (1909). *The arts & crafts of ancient Egypt.* London, England: Bracken.

Petry, A. (1946). *The street.* Boston: Houghton Mifflin.

Petry, A. (1947). *Country place.* Boston: Houghton Mifflin.

Petry, A. (1953). *The narrows.* Boston: Houghton Mifflin.

Phillips, F. B. (1996). NTU psychotherapy: Principles and processes. In D. A. A. Azibo (Ed.), *African psychology in historical perspective & related commentary* (pp. 83-97). Trenton, NJ: Africa World Press.

Phillips, F. B. (1998). Spirit-energy and NTU psychotherapy. In R. L. Jones (Ed.), *African American mental health: Theory, research, and intervention,* (pp. 357-377). Hampton, VA: Cobb &

Henry.

Prilleltensky, I., & Gonick, L. S. (1994). The discourse of oppression in the social sciences: Past, present, and future. In E. J. Trickett, R. J. Watts, & D. Birman (Eds.), *Human diversity: Perspectives on people in context* (pp. 145-169). San Francisco, CA: Jossey-Bass.

Radin, P. (Ed.). (1952/1983). *African folktales.* New York: Schocken.

Rashidi, R. (1992). *Introduction to the study of African classical civilizations.* London, England: Karnak House.

Rathele, V. N. & Mutwa, C. (2008). *Woman of four paths: The strange story of a Black woman in South Africa.* Kuruman, South Africa: Tso-Dilo Media.

Rhines, J. A. (1996). *Black film/White money.* New Brunswick, NJ: Rutgers.

Robinson, W. H. (Ed.). (1971). *Early Black American prose.* Dubuque, IA: William C. Brown.

Rodgers, L. R. (1997). *Canaan bound: The African-American great migration novel.* Urbana, IL: University of Illinois.

Rodney, W. (1972). *How Europe underdeveloped Africa.* Washington, DC: Howard.

Rodriguez, J. (2003, February). *A psychologist's view on hip-hop lyrics.* Retrieved on December 19, 2005, from http//www.allhiphop. com/features/?ID'514

Rosenblatt, R. (1974). *Black fiction.* Cambridge, MA: Harvard.

Rowe, D. M. (1995). The nommotization of the African Psychology Institute. *Psych Discourse, 26* (7), 10.

Rowe, T. D., & Webb-Msemaji, F. (2004). African-centered psychology in the community. In R. L. Jones (Ed.), *Black psychology* (4th ed., pp. 701-721). Hampton, VA: Cobb & Henry.

Rowell, C. H. (1985). An interview with Larry Neal. *Callaloo. 8*:1,

11-35.

Saluting all-time greatest jockey. (2005, August 1). USA Today (Magazine). Retrieved on December 26, 2005, from http://highbeam.com/library/docfree.asp?DOCID'1G1:35076832&ctrlInfo'Roun

Sanders-Thompson, V. L. (1995). The multidimensional structure of racial identification. *Journal of Research in Personality, 29*(2), 208-222.

Savery, P. (1989). "Not like an arrow, but a boomerang": Ellison's existentialblues. In S. R. Parr, & P. Savery (Eds.), *Approaches to Teaching Ellison's Invisible Man* (pp. 65-74). New York: The Modern Language Association of America.

Schafer, W. (1970). Irony From Underground-Satiric Elements in *Invisible Man.*In. J. M. Reilly (Ed.), *Twentieth century interpretations of Invisible Man* (pp. 39-47). Englewood Cliffs, NJ: Prentice-Hall.

Schraufnagel, N. (1973). *From apology to protest: the Black American novel.* Deland, FL: Everett/Edwards.

Scruggs, C. (1993). *Sweet home: Invisible cities in the Afro-American novel.* Baltimore: Johns Hopkins.

Seale, B. (1968). *Seize the time: The story of the Black Panther Party and Huey P. Newton.* New York: Vintage.

Sellers, R. M., Chavous, T. M., & Cooke, D. Y. (1998). Racial ideology and racial centrality as predictors of African American college students' academic performance. *Journal of Black Psychology, 24*(1), 8-27.

Semaj, L. T. (1981). The black self, identity, and models for a psychology of black liberation. *Western Journal of Black Studies, 5*, 158-171.

Serequeberhan, T. (1991). *African philosophy: The essential readings.*

New York: Paragon House.

Sexton-Radek, K. (Ed.). (2005). *Violence in schools: Issues, consequences, and expressions*: Westport, CT: Praeger.

Sizemore, B. A. (1972). Social science and education for a Black identity. In J. A. Banks, & J. D. Grambs (Eds.), *Black self-concept: Implications for education and social science* (pp. 141-170). New York: McGraw-Hill.

Slatoff, W. (1989). Making *Invisible Man* matter. In S. R. Parr, & P. Savery, (Eds.), *Approaches to Teaching Ellison's Invisible Man* (pp. 31-36). New York: The Modern Language Association of America.

Smallwood, C., & Shields, G. (1997). *Quasi-morticide: Self destructive behavior: Reversing the cycle in the African-American community.* Pittsburgh, PA: Dorrance Publishing.

Smedley, A. (1999). *Race in North America: Origin and evolution of aworldview.* Boulder, CO: Westview.

Smedley, A., & Smedley, B. D. (2005). Race as biology is fiction, racism as a social problem is real. *American Psychologist, 60*(1), 16-26.

Smith, R., & Jones, S. L. (Eds.). (2000). *The Prentice Hall anthology of African American literature.* Upper Saddle River, NJ: Prentice Hall.

Sollors, W. (1997). *Neither Black nor White yet both: Thematic explorations of interracial literature.* Cambridge, MA: Harvard.

Sollors, W. (Ed.). (2000). *Interracialism: Black-White intermarriage in American history, literature, and law.* New York: Oxford.

Some, M. P. (1993). *Ritual: Power, healing, and community.* New York: Penguin. Some, M. P. (1998). *The healing wisdom of Africa: Finding life purpose through nature, ritual and community.* New York: Jeremy P. Tarcher/Putnam.

Some, S. E. (2003). *Falling out of grace: Meditations on loss, healing and*

Wisdom. El Sobrante, CA: North Bay Books.

Spark, C. L. (2001). *Hunting Captain Ahab: Psychological warfare and the Melville revival*. Kent, OH: Kent State University.

Ssekitooleko, D. (2005). *The challenge of African humanism*. Retrieved on May 11, 2005, from http://www.secularhumanism.org/library/aah/ssekitooleko_12_1.html

Ssensalo, B. M. (1978). *The Black psudo-autobiographical novel: A descendant of Black autobiography*. Unpublished doctoral dissertation, University of California, Los Angeles.

Staples, R. (1982). *Black masculinity: The Black male's role in American society*. San Francisco: The Black Scholar Press.

Starke, C. J. (1971). *Black portraiture in American fiction: Stock characters, archetypes, and individuals*. New York: Basic.

Stephens, G. (1999). *On racial frontiers: The new culture of Frederick Douglass, Ralph Ellison, and Bob Marley*. New York: Cambridge.

Stewart, J. (1996). *1,001 African names: First and last names from the African continent*. Sacramento, CA: Citadel Press.

Strout, C. (1989). "An American Negro Idiom"; *Invisible Man* and the politics of culture. In S. R. Parr, & P. Savery (Eds.), *Approaches to Teaching Ellison's Invisible Man* (pp. 79-85). New York: The Modern Language Association of America.

Sutherland, M. (1993). *Black authenticity: A psychology for liberating people of African descent*. Chicago: Third World Press.

Suzar. (1999). *Blacked out through whitewash*. Oak View, CA: A-Kar Productions.

Swahili name book. (1971). NewArk, NJ: Jihad.

Tanner, T. (1974). The music of invisibility. In J. Hersey (Ed.). *Ralph Ellison: A collection of critical essays* (pp. 80-94). Englewood Cliffs, NJ: Prentice-Hall.

Tate, C. (1987). Notes on the Invisible Women in Ralph Ellison's *Invisible Man*. In K. W. Benston (Ed.), *Speaking for you: The vision of Ralph Ellison* (pp. 163-171). Washington, D.C.: Howard University.

Tate, C. (1998). *Psychoanalysis and black novels: Desire and the protocols of race*. New York, NY: Oxford University Press.

Taylor, J. (1998). Cultural conversion experiences: Implications for mental health research and treatment. In R. L. Jones (Ed.), *African American identity development* (pp. 85-97). Hampton, VA: Cobb & Henry.

Tetteh, I. N. O. (1997). *Soul processing: The path to freedom.* Accra-North, Ghana: The Etherean Mission.

Thomas, C. W. (1971). *Boys no more.* Beverly Hills, CA: Glencoe.

Thomas, A., & Sillen, S. (1993). *Racism & psychiatry*. New York: Citadel. Thornton, J. (1993). Central African names and African-American naming patterns. *The William and Mary Quarterly, 3rd Series, L*(4), 727-742.

Thrall, W. F., Hibbard, A, & Holman, C. H. (1960). *A handbook to literature.* New York: Odyssey.

Three Initiates. (1940). *The Kybalion: A study of the Hermetic philosophy of ancient Egypt and Greece*. Chicago: The Yogi Publication Society.

Toure, A. S. (1973). *Afrika and imperialism.* NewArk, NJ: Jihad.

Tucker, C. M., & Herman, K. C. (2002). Using culturally sensitive theories and research to meet the academic needs of low-income African American children. *American Psychologist, 57*(10), 762-773.

Tyler, K. M., Boykin, A. W., Boelter, C. A., & Dillihunt, M. L. (2005). Examining mainstream and Afro-cultural value socialization in African American households. *Journal of Black Psychology,*

31(3), 291-310.

Vogler, T. A. (1974). Invisible Man: Somebody's Protest Novel. In J. Hersey (Ed.), *Ralph Ellison: A collection of critical essays* (pp. 127-150). Englewood Cliffs, NJ: Prentice-Hall.

Wall, C. A. (1994). On freedom and the will to adorn. In G. Levine (Ed.), *Aesthetics and ideology* (pp. 283-303). New Brunswick, NJ: Rutgers.

Wallace, M. O. (2002). *Constructing the Black masculine: Identity and ideality in African American men's literature and culture, 1775-1995.* Durham, NC: Duke.

Walling, W. (1973). Ralph Ellison's *Invisible Man*: "It goes a long way back, some twenty years". *Phylon, 34*(1), 4-16.

Warren, K. W. (2003). *So black and blue: Ralph Ellison and the occasion of criticism.* University of Chicago.

Warren, R. P. (1974). The Unity of Experience. In J. Hersey (Ed.), *Ralph Ellison: A collection of critical essays* (pp. 21- 26). Englewood Cliffs, NJ: Prentice-Hall.

Washington, B. T. (1906). *Up from slavery: An autobiography.* New York: Doubleday, Page.

Watts, R. J., Griffith, D. M., & Abdul-Adil, J. (1999). Sociopolitical development as an antidote for oppression - theory and action. *American Journal of Community Psychology, 27*(2), 255-271.

Way, N., & Chu, J. Y. (Eds.). (2004). *Adolescent boys: Exploring diverse cultures of boyhood.* New York University.

Webb-Msemaji, R. (1996). *The impact of African self-consciousness, ethnic identity, and racial mistrust on the self-esteem of African American adolescents.* Unpublished doctoral dissertation, California School of Professional Psychology, Los Angeles.

Wells-Barnett, I. B. (1969). *On lynchings.* New York: Arno Press.

Welsing, F. C. (1991). *The Isis papers: The keys to the colors.* Washington,

DC: C.W. Publishing.

West. A. (1970). Black man's burden. In J. M. Reilly (Ed.), *Twentieth century interpretations of Invisible Man* (pp. 102-106). Englewood Cliffs, NJ: Prentice-Hall.

White, J. L. (1970). Toward a Black psychology. *The Black Scholar, 1*(5), 52-57.

White, J. L., & Parham, T. A. (1990). *The psychology of Blacks* (2nd ed.). Englewood Cliffs, NJ: Prentice Hall.

White, W. (1969). *Rope and Faggot: A biography of Judge Lynch*. New York: Arno.

Wilber, R. E. M. (1999). The dearth in African-American literature: A social psychological analysis of the missing fiction genres and the correlation to the major social, economic, and mental health problems of the African-American community and the effects on television programming. *Dissertation Abstracts International, 59*(09), 3460. (UMI No. AAT 9907575)

Wilcox, R. (1971). Reprise: In the light of increased awareness. In R. Wilcox (Ed.), *The psychological consequences of being a Black American* (pp. 477-479). New York: John Wiley.

Williams, C. (1987). *The destruction of Black civilization: Great issues of a race from 4500 b.c. to 2000 a.d.* Chicago, IL: Third World.

Williams, R. L. (1981). The collective black mind: An Afrocentric theory of black personality. St. Louis, MO: Williams and Associates.

Williamson, J. (1984). *New people: Miscegenation and mulattoes in the United States*. New York University.

Wilson, A. N. (1993). *The falsification of Afrikan consciousness: Eurocentric history, psychiatry and politics of white supremacy*. New York: Afrikan World InfoSystems.

Wilson, A. N. (1999). *Afrikan-centered consciousness versus the new world order: Garveyism in the age of globalism.* New York: Afrikan World InfoSystems.

Wilson, H. E. (1859/1983). *Our nig; or, sketches from the life of a free Black.* New York: Vintage.

Wimby, R. (1986). The unity of African languages. In M. Karenga & J. H. Carruthers (Eds.), *Kemet and the African worldview: Research, rescue and restoration* (pp. 151-166). Los Angeles: University of Sankore.

Wise, T. (1999, April 27) . Famous last words: Exploring the depths of racist conditioning. *Znet commentary.* Retrieved on September 18, 2001, from http://www.fiskrii.org/articles-essays/wise/tolerance.html

Wise, T. (2005). *White like me: Reflections on race from a privileged son.* Brooklyn, NY: Soft Skull Press.

Wolfenstein, E. V. (2003). Race, rage, and Oedipus in Ralph Ellison's *Invisible Man.* In D. Moss (Ed.), *Hating in the first person personal: Psychoanalytic essays on racism, homophobia, misogyny, and terror* (pp. 69-113. New York: Other Press.

Woodson, C. G. (1933). *Miseducation of the Negro.* Washington, D.C.: The Associated Publishers.

Woodward, C. V. (1974). *The strange career of jim crow* (3rd ed.). New York: Oxford.

Wright, B. E. (1984). The psychopathic racial personality and other essays.Chicago: Third World.

Wright, R. (1936/1965). *Uncle Tom's children.* New York: Harper & Row.

Wright, R. (1937). *Black boy.* New York: Harper.

Wright, R. (1937/1978). Blueprint for Negro writing. In E. Wright & M. Fabre (Eds.). *Richard Wright reader.* (Pp. 37-49). New York: Harper & Row.

Wright, R. (1940/1966). *Native son.* New York: Harper & Row.

Wright, R. (1941/1988). *12 million Black voices.* New York: Thunder's Mouth.

Wright, R. (1944/1969). The man who lived underground. In R. Wright (Ed.). *Eight men,* (pp. 22-74). New York: Pyramid.

Wright, R. (1944/1977). *American hunger.* New York: Harper & Row. Wright, R. (1954). *Black power.* New York: Harper & Brothers.

www.ingramcontent.com/pod-product-compliance
Lightning Source LLC
Chambersburg PA
CBHW061348280526
45784CB00001B/188